BIM Strategic Implementation: Benchmarking Based Decision Making Framework

Dr. Keyu CHEN

Dr. Haijiang LI

Copyright © 2017 Keyu CHEN

All rights reserved.

ISBN: 1973904438

ISBN-13: 978-1973904434

To my parents, friends,

For their understanding,

encouragement, support and love.

ACKNOWLEDGMENTS

The author would like to express his special thanks to:

Dr. Haijiang Li

Professor Yacine Rezgui

Dr. Alan Kwan

Professor Benachir Medjdoub

who gave him support and advice on both professional knowledge and for helping to pave his future career;

The author also greatly appreciates the help from his industry partners:

Professor Jeff Perren

Mr. Jason Jones

Mr. Haichuan Dong

Dr. Michael Loh

for their knowledge sharing and assistance throughout his work;

And finally:

Mr. Derek Murray

Mr. Nicholas Nisbet

for their review and providing forewords for this book;

In addition to the above, a big thanks to those who have provided the author with support and help along this journey.

PREFACE

Building easily outlast their inhabitants. Buildings last forever while civilizations rise and fall. Edifices remain as testaments of cultures and history, other than dwelling for humans. It also represents the advanced level of technology and human society. Building Information Modelling (BIM) in today's building industry is believed as a holistic evolution. It started with software breakthrough. After a while the practitioners realized a proper BIM implementation specification with a focus on the technology, data, process, legal, people and culture is needed to assist BIM users to achieve a standardized adoption procedure and better results. Scholars furthermore argue that it is not necessary to adopt full BIM, instead, a lean BIM concept should be considered.

This book introduced a benchmark for BIM users to assess their current BIM capability, strength and weakness in project, organization and industry level. Hence, the most informed decision and strategic plan can be made. Two vital elements of this framework are: a comprehensive set of decision making criteria and a reasonable priority system – where the weightage assigned to each criteria should align with the company's objective and vision. The proposed concept can be applied regardless of project types, regions, regulations, clients' requirements etc. while specific information could definitely improve the accuracy of the framework and achieve a much higher level of BIM.

This book is for high end BIM users and guide their daily BIM tasks. Those users include but are not limited to: BIM managers, regional BIM directors, BIM strategist, project managers from all parties including designer, contractor, QS, developers as well as the facility management group. This book is also suitable for BIM modelers, coordinators, designers, builders and other individuals who are engaged in a BIM based project and would like to go further.

This book has demonstrated the proposed benchmark, its unique development method, application in real projects and lesson learned, take away points for reader's better understanding.

FOREWORD BY DEREK

By reading this book you have already made the first of many good decisions; you are a step closer to making real positive change in advancing yourself and our Architecture, Engineering and Construction (AEC) industry. This book will provide you with a framework to prioritise your next decision on the BIM journey – well done and welcome aboard!

Dr. Chen has experienced the emerging, global world of BIM first-hand, across the technologies, processes and industry cultures that influence our collective progress and asset performance outcomes. His work in the UK helped shape how organizations set out appropriate strategies for success, backed with benchmarking approaches to support advancement. This was during a key time for the UK construction industry, with a government mandate driving industry change in project delivery to achieve significant reduction in the capital and operational costs of built assets. His passion and enthusiasm for changing the way we work has now switched to Singapore, in leading the use of BIM to bring performance improvements across the AEC industry; all of which will contribute to a 'smart nation' vision.

This publication is essential in meeting not only Singapore's vision but in advancing all industry players, irrespective of specialty or geography. Without a means of making the most appropriate decisions in our rapidly changing technological landscape the AEC industry will struggle to advance. It's accepted globally that it is no longer a question of 'whether' AEC will be disrupted, but 'when'.

In today's context, we have to design, construct and operate our assets with increasingly complicated and diverse requirements. Ensuring appropriate decisions are made from the very outset of a project has the biggest influence on project outcomes across safety, sustainability, resilience and adaptability of use. Retrospective decision making is costly and those costs can rise beyond control. And as time goes on it becomes increasingly hard to influence outcomes.

As Global Head of Digital Projects at Mott MacDonald I have experienced a diverse range of industries and geographies on their journeys to success. Working with our global BIM Practice, we have recognised the need for appropriate decision frameworks in helping move the needle when it comes to internal efficiency, delivering differently and providing our clients with successful outcomes.

Since 2010, Mott MacDonald has had a formal BIM strategy lead from the very top of the organization. It has been refreshed twice to reflect the changing nature of industry and the decisions we now make in working with our clients and the supply chain. We have shifted focus from early adoption of BIM process and tools and developing our teams, to asset management and a world of smart infrastructure where real world sensors and data are abundant. This is all supported by a global maturity assessment framework facilitating continuous improvement and healthy competition between the connected parts of our organization.

The BIM strategy has established a number of enablers to facilitate widespread BIM adoption and remove barriers to making decisions, including:

- A Common Data Environment (CDE) mapped on to BS 1192 and delivered via a global technology platform

- A Digital Component Catalogue (DCC) providing consistent, information-rich objects for our projects

- Enhanced enterprise agreements with key technology partners for software and learning

- A specialist BIM Consultancy team to assist clients in making the right decisions on their BIM journey

- A global network of digital delivery leaders, supporting each other to advance the way we work

- A knowledge and communication platform which connects our teams in formal and social contexts

- An overall enhanced company skill base, with learning tools and assessments supporting development

The technologies we now apply are supported by decisions made at the very outset of a project. Where our client is mature in specifying information requirements our job is easier and we are able to turn our attention to innovation in improving not only design outcomes, but those in construction and operations. The winners are the asset owners and those who use those assets. In less mature environments, we work with our clients to help them understand the benefits of information development and how they can apply appropriate technologies in their advancement. We implement appropriate strategies and decision making frameworks with our clients and undertake early pilots to demonstrate successful adoption.

More recently, advances in mixed realities have been allowing us to innovate with early BIM adopters. Our decision to adopt a consistent global implementation of virtual reality has provided a rapid upscaling of the technology and ability to share use cases globally, taking virtual reality beyond a simple game into a collaborative experience for our distributed design teams and our clients.

As we become more connected through technology, the ease at which we can share information is also a risk. Decisions around information security are now at the forefront of asset owners' agendas. It's no longer acceptable to just 'send an attachment' and all projects should consider appropriate levels of confidentiality, integrity and availability in their decision making. By taking such considerations, it will be possible to make practical gains in terms of rapid access to the correct version of asset information, no matter what stage of the project.

Of course, the ultimate test is to measure productivity output. Without any measures of BIM success this simply will not be possible. In Singapore, we are fortunate to have some of the best infrastructure developments to meet our challenging environment and densely populated demands. However, we still observe a need to rapidly up the game in terms of BIM adoption. On a recent project, we decided to introduce novel information-rich technology to a contractor organization, it was simply the 'right thing to do'. Through a partnering-style approach of joint workshops and early pilots, we have gained the confidence of the contractor, and are together celebrating the best approach they have seen to digital delivery. Of course, the real beneficiaries are those end users, the taxpayers who consume services. Better industry decisions lead to better value per dollar. Not all cases are successful, and resistance to change elsewhere has hamstrung our organization and the supply chain to continue delivering in the same way we did 10 years ago. We continue to address cases like this and are steadily winning over even the most sceptical stakeholders. Overall though, AEC is sadly behind the curve when it comes to being digital. At the root of this is having appropriate decision making in place on which to build confidence, maturity and successful outcomes on projects.

Different countries and industries have adopted BIM through slightly different views through the lens and Dr Chen's work provides a means to adopt a pragmatic approach to decision making. In some countries — such as the United Kingdom — a central government approach in focusing BIM on process across the entire lifecycle of assets will contribute to overall reduction in procurement and operational costs thanks to all round better information. In Singapore, the leading focus has been on the design phase of assets, resulting in greater accuracy for building approvals. Shift to the construction phase should now help the wider supply chain with adoption. And with appropriate decision making in place we will hopefully see this move towards asset management and operations sooner rather than later.

At Mott MacDonald we look forward to learning from Dr Chen's work in informing and further advancing our own decision making and existing, global BIM maturity assessment framework. We will continue to drive leading-edge innovation in the application of BIM such that we are part of the inevitable disruption. I encourage all organizations, across the entire AEC supply chain, to do the same, be there with us and make a positive difference to your project outcomes starting today.

Derek Murray, MEng CEng MICE, is Mott MacDonald's Global Head of Digital Projects, leading the organization's GoDigital Transformation programme across technology innovation, automation and digital delivery components. Now operating from Singapore, Derek has 25 years' experience in the civil engineering industry, he has previously served global and regional technology leadership roles in the UK, Middle East and South-East Asia.

FOREWORD BY NICHOLAS

Forty years after the first commercial BIM applications emerged in the UK, BIM has now become business as usual in the UK and in most developed countries around the world. Even those who decline to adopt it know that their choice is perverse and risky. In 2009 AEC3 prepared 'The Business Case for BIM' for the UK Government and its value was shown the following year when this was translated into the hypothesis 'The [UK] Government as a client can obtain improved cost, value and carbon from the use of shared structured information'. We contributed to the industry response, and helped get the BS 1192 series of standards developed. The UK Government Mandate to its central spending departments has now created the momentum that has ensured that the UK Standards and policies are being studied, replicated and proved around the world. Now BIM is evolving and integrating to become the flagship for the total digitisation of the built environment. The empowerment of Smart Communities, the exploitation of the Internet of Things and the connection to social outcomes all depend on the core information that BIM creates and manages. These will grow the challenges associated with data management and interoperability.

One question remains: can the industry respond to these challenges? Around the world many firms are still being led by their clients, contracts and obligations. But this 'pull' must now be matched by the industry beginning to 'push' itself forward with purposeful, well considered aims. The urgent need is to actively engage with not only those demands but also with the opportunities for better processes and outcomes. Every organisation, but especially those in design and engineering, need to measure and manage their standing and their progress on organizational, data, project, applications and stakeholder involvement.

There is a compelling need to take a holistic approach which can evolve and adapt year-on-year. Dr. Chen Keyu has outlined a clear and sensible approach, focusing on distinct, recognizable criteria, and the achievement of balance between them. Directors, managers and consultants charged with ensuring the sustainability of their firms, and the clients' interests will want to use the careful assessments outlined here to monitor and steer a course between on the one hand commercial hype and on the other hand debilitating caution.

If BIM is about anyone thing, it is about making information about the built environment re-usable and comparable: the assessment and benchmarking of our competency is a vital tool.

Nicholas Nisbet, MA (Cantab) DipArch (UNL), Director and Owner of AEC3 Ltd, Vice-Chair of buildingSMART UK Chapter, ifcXML Coordinator of buildingSMART Model Support Group.

In London, 2017

CONTENTS

ACKNOWLEDGMENTS ii
PREFACE iii
FOREWORD BY DEREK v
FOREWORD BY NICHOLAS xi
CONTENTS xiii
LIST OF FIGURES xv
LIST OF TABLES xvi
GLOSSARY 1
DISCLAIMER 2
1 INTRODUCTION 1
   1.1 Background 1
   1.2 Motivation 3
   1.3 Aim and objectives 8
   1.4 Outline 8
2 LITERATURE REVIEW 10
   2.1 BIM introduction 10
   2.2 BIM dimensions 13
   2.3 BIM implementation 15
      2.3.1 BIM standards, guidelines, executions plans and pro-tocols etc. 16
      2.3.2 BIM Implementation Framework 20
   2.4 BIM Assessment and Benchmarking 24
      2.4.1 BIM Evaluation & assessment 24
      2.4.2 BIM Project Benchmarking 31
   2.5 Summary 35
3 FRAMEWORK DEVELOPMENT 37
   3.1 Framework Design 37
   3.2 Research context 38
   3.3 Framework refinement 39
   3.4 Priority and sensitivity analysis 40
   3.5 Framework validation and application 44

3.6 Summary 46
4 DIMENSIONS & FACTORS FOR BIM IMPLEMENTA-TION AND EVALUATION 48
4.1 Preliminary BIM implementation Framework 48
4.2 Framework enhancement through practical case study ... 50
    4.2.1 BIM pilot project 50
        4.2.1.1 Collaboration environment set up 50
        4.2.1.2 Practical adoption issues and solutions 52
    4.2.2 Real-life industry case study 53
        4.2.2.1 Project introduction 53
        4.2.2.2 Case study design 54
        4.2.2.3 Observation for existing practices 55
        4.2.2.4 BIM strategy planning and implementation 58
        4.2.2.5 Result and feedback 63
    4.2.3 Framework revise based on expert consensus 65
4.3 Summary 70

5 AHP BASED STRATEGIC DECISION MAKING FOR ORGANIZATIONAL BIM IMPLEMENTATION 71
5.1 Analytical hierarchy model development 71
5.2 The weighting system development 75

6 BIM EVALUATION FRAMEWORK (BEF) DEVELOP-MENT & VALIDATION 78
6.1 BIM evaluation Framework (BeF) development 78
6.2 Validation result & analysis 84
    6.2.1 Comparison among practical projects 84
    6.2.2 Comparison with existing methods 92
    6.2.3 User's feedback 93
6.3 Sensitivity analysis 95
    6.3.1 Ranking score of focus schemes 95
    6.3.2 New priorities & strategy 96
6.4 Summary 101

7 CONCLUSION 103
REFERENCES 106
POSTSCRIPT 119

LIST OF FIGURES

Figure 1-1 Construction & non-Farm Labor Productivity Index (1964-2003) Source: US Dept. of Commerce, Bureau of Labor Statistics   2

Figure 2-1 BIM maturity levels by Bew –Richards (BIMIWG, 2011)   11

Figure 4-1 The Preliminary BIM implementation framework   47

Figure 4-2 The proposed BIM implementation framework by Delphi study   66

Figure 5-1 Hierarchy model for BIM implementation in ASZ   70

Figure 6-1 Example of proposed BIM evaluation Framework (BeF)   80

Figure 6-2 Project A-E results comparison between existing methods' average value and BeF   90

Figure 6-3 Example of sensitivity analysis: impact of Sustainability towards Focus schemes   94

Figure 6-4 Comparison of alternatives rankings before and after factor priorities shifting   98

LIST OF TABLES

Table 4-1 Project Milestones of real project in SAIXA   59

Table 4-2 Performance comparison between BIM group and traditional group in case study in SAIXA   62-63

Table 5-1 Result of AHP study: global and local priority and CR value   71-72

Table 6-1 Validation result of project A-E   81

Table 6-2 nD BIM model comparison among project A-E   83

Table 6-3 Key Performance Indicators for project A-E   84

Table 6-4 Project performance comparison for project A-E   86

Table 6-5 Project A-E results comparison between existing methods' average value and BeF   90

Table 6-6 Ranking order of future objectives & contribution from each dimension   92

Table 6-7 Example of sensitivity analysis: impact of *Sustainability* towards Focus schemes   93

Table 6-8 Sensitivity analysis of factor & dimension change trend   95

Table 6-9 Global priority before and after Sensitivity analysis   96-97

GLOSSARY

AEC	Architectural Engineering and Construction
AHP	Analytic Hierarchy Process
BeF	BIM evaluation Framework
BiF	BIM implementation Framework
BIM	Building Information Modelling
BPM	BIM proficiency Matrix
BREEAM	Building Research Establishment Environmental Assessment Methodology
CAD	Computer Aided Design
CIC	Computer Integrated Construction
CIFE	Centre for Integrated Facility Engineering
CR	Consistency Ratio
FM	Facility Management
I-CMM	Interactive Capability Maturity Model
IQR	Interquartile range
IS	Information System
KPI	Key Performance Indicator
LEED	Leadership in Energy & Environmental Design
LOD	Level of Development
MCDM	Multiple Criteria Decision Making
MEP	Mechanical, electrical, and plumbing services
NBS	National BIM Specification
PM	Project Management
ROI	Return on Investment
SME	Small and Medium Entrepreneur
VDC	Virtual Design and Construction
4D BIM	Four Dimensional BIM
5D BIM	Five Dimensional BIM

DISCLAIMER

This book was developed in the author's personal capacity and in no way represents Building and Construction Authority Singapore's (BCA) views or endorsement.

1 INTRODUCTION

1.1 Background

The construction industry is seen as one of the most challenging and complex industries in many countries. The UK government admits that the construction industry is lagging behind other industries in terms of fully utilizing digital technology (Cabinet Office, 2011). The reasons are (Porwal and Hewage, 2013, Soares, 2013): (1) the culture in the AEC industry makes it difficult for information to be reused and shared; (2) a majority of the information during the project lifecycle is not re-usable, while poor information management leads to unnecessary duplication of data, project delay and budget overruns; (3) and the entire work flow is fragmented and there is a lack of effective communication between partners. It is difficult to manage change orders, design, cost estimates or planning, and there are big gaps between the design process and the construction activities.

All these lead to low productivity in AEC organizations for a long time. A study by the Centre for Integrated Facility Engineering at Stanford University found that the productivity of the Architecture, Engineering and Construction (AEC) industry has decreased by more than 10% from 1964-2003 (Figure 1-1) (according to the U.S. Department of Commerce, Bureau of Labor statistics). By contrast during the same period, many other sectors have had a dramatic increase in productivity (Adriaanse et al., 2010). One of the main consequences of low productivity is the waste of resources (Eastman et al., 2011, Soares, 2013).

Building Information Modelling (BIM) is a recently emerged digital concept in AEC industry and it is believed that it can help improve productivity as well as have many other benefits (Rezgui and Medjdoub, 2007, Succar, 2009a, Eastman et al., 2011). They work towards a better project performance and outcome (Wang et al., 2015). It has been found that even during the recent economic recession, UK AEC industry is still willing to promote BIM (Cabinet Office, 2011).

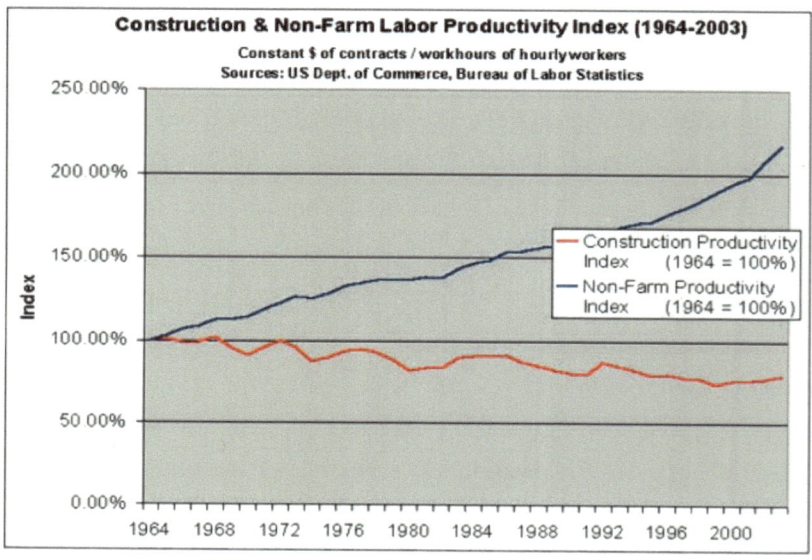

Figure 1-1 Construction & non-Farm Labor Productivity Index (1964-2003) Source: US Dept. of Commerce, Bureau of Labor Statistics

However, in BIM practice, the transformation from traditional 2D based approach to collaborative BIM, and adherence to legislation (such as building regulations, environment requirements and local government agencies) in current AEC projects is still low, which is well below the industry's expectation (NBS, 2014, McGraw Hill Construction, 2010).

A survey conducted by the National BIM Specification (NBS) (NBS, 2014) shows the number of BIM users has an increasing rate, which has increased from 13% to 54% (of the total participant) in year 2010 and 2013. Nearly 50% of respondents hold a negative opinion regarding the future of BIM due to: a lack of confidence in their BIM skill; lack of awareness, knowledge and investment priorities on how to meet minimum BIM requirements. Therefore, there are existing overestimation of BIM implementation cost delay and hesitation on the uptake of BIM (Aranda-Mena et al., 2009, Forsythe, 2014, Mom et al., 2014b). A survey conducted by SmartMarket revealed that there are gaps on BIM assessment, functionality selection and client demands for current BIM implementation (McGraw Hill Construction, 2010, Won et al., 2013). BIM was believed to add extra new and complex technologies on top of those already complex and fragmented traditional design and construction process; All these have made practical BIM implementation very difficult and convoluted (Chien et al., 2014).

The integration and merging of the most recognized implementation methodologies could provide a unified awareness, knowledge and understanding among all stakeholders within BIM implementation (Succar, 2009a, Kam et al., 2013a, Forsythe, 2014). These implementation methodologies can also be the key to achieving a more matured BIM standard, and applied by decision makers as a basis to develop a comprehensive BIM execution plan at the commencement of the project (AIA, 2007, Khosrowshahi and Arayici, 2012).

1.2 Motivation

With the development of BIM, the main focus has been shifted from technology (Brynjolfsson, 1993, Jung and Gibson, 1999) to include management, process, people and policy etc., and that has greatly increased the complexity of BIM implementation process (Succar, 2009a). AEC industry contains multiple activities in various stages and involves participants from different sectors/disciplines (Jung and Gibson, 1999). This partnership could be easily interfered due to different backgrounds, disciplines and geographical locations as well as representing various interests, benefits, risks, competencies, and maturity levels (Gu and London, 2010, Khosrowshahi and Arayici, 2012).

Existing guidelines tailored for different parties, adoption philosophies, technologies required, expected profits, implementation roadmaps and methods of risk management etc. are all different (Gao, 2011, Chien et al., 2014). It is therefore believed that the more parties involved, the wider range of knowledge covered, the more difficult the decision making process will be (Wang et al., 2005). In developing countries e.g. China, after the economic reform, more nationwide partners have been involved in a single project which has greatly increased the complexity of the project, and the difficulty of the evaluation process. BIM users also have a variety of experience, knowledge and understanding in terms of practical BIM implementation which causes different levels of adoption (Eadie et al., 2013). All these lead to an inconsistency of BIM usage level, performance and adoption, even within the same project. Without a unified standard for BIM implementation, diverse of methods, maturity and compliance levels of BIM among project partners will be happened, which can cause interoperability problems (Giel et al., 2012).

Lack of priority leads to high adoption cost

Succar (2010) argues that there is a lack of guidance to prioritize requirements in order to enhance BIM deliverables, this increases risks and budgets during the implementation process (Aranda-Mena et al., 2009, Sebastian and van Berlo, 2010). Porwal and Hewage argues the initial adoption of BIM in a construction project requires significant resources, such as long term training; upgrade of hardware and software license purchases; engineering analysis and simulation activities etc. (Porwal and Hewage, 2013). Moreover, with BIM the AEC industry is now looking for both commercial value and sustainability certification. Due to the lack of adoption priorities, such a full adoption of BIM leads to an early cost increase instead of saving, which causes delay and hesitation of a BIM uptake (Rezgui and Medjdoub, 2007, Won et al., 2013).

The use of BIM in large corporations is also very difficult

(Khosrowshahi and Arayici, 2012). While Small and Medium Entrepreneur (SME's) have to be more careful with their resources, large corporations normally have a more matured processes and management approach. The transformation to BIM normally means the change of everything: employees' skill, hardware, software, business process and new services etc. (Khosrowshahi and Arayici, 2012). This leads to a great risk as the actual return on investment could be lower than the expected. Before the actual transformation, a series of complex feasibility studies, Return on Investments (ROI) calculations and business case proposals should be approved by the board of directors. While these efforts only provide an evaluation of an optimized outcome, there are still lack of guidance on how to achieve this, especially in an efficiency way.

Therefore, in order to balance resource and trade-off investment, a unified weightage system is needed (Sebastian and van Berlo, 2010, Jeong et al., 2013). This will assist decision making process of different project parties regardless of their roles and scales, to clarify a unified requirement, target, prioritized key focus area and objectives' which will benefit and assist BIM adoption. This will also reduce its risk and speed up its initial assessment process.

Defects of existing BIM assessment framework

In order to decide what to do or what is the most urgent, the intuition or past experience based implementation approach is not sufficient (Gao, 2011). Bloom and Reenen (2006) believe improvement in terms of productivity, profitability and business opportunities can be made by assessing and improving the management process, which also improves project performance (Kam et al., 2013a, Miettinen and Paavola, 2014). It is also one of the most effective ways to determine the organization's real-time performance of BIM, as well as what is to be aligned with the BIM 'vision', use of BIM and to achieve a desired maturity level within the organization's BIM strategy plan (Sebastian and van Berlo, 2010, Du et al., 2014, PSCIC, 2013, NIBS,

2007b). Barry et al. (2012) believed it is necessary to carry out an initial assessment to identify what are the most needed implementation areas when considering BIM, especially when financial resources are tight. Porwal et al. (2013) also agree assessing the organization's capability will help new BIM users to get started. Moreover, such an evaluation method is appropriate for individuals (e.g. BIM manager). It is expected to be used prior to the commencement of a AEC project (Wang et al., 2005). Therefore, the strategy for the use of BIM in the project, with goals and paths for a better project outcome, can be developed by the decision maker.

Even though there are a number of BIM evaluation / assessment methods available such as (NIBS, 2007b, Kreider, 2011, Indiana University, 2009b, Duncan and Aldwinckle, 2015) etc., they have not been applied consistently in reality, due to limitations in term of: assessment criteria selection, criteria weightage calculation, assessment coverage, outdated/expandability, validation and impact of criteria towards organization's goal etc.

Hence, there is a need to develop a more tangible and practical evaluation framework which underlying concept of BIM regardless of the different requirements and types of project etc. most importantly, to consider the objectives of the organization and which contains all dimensions and factors relevant to a BIM implementation verified by all stakeholders with a consensus (Chen et al., 2014).

Design stage problems and its potential solution

More project pressure and workload has been allocated to the design stage: the richness and accuracy of the nD BIM model is essential during this stage, information relevant to design, fabrication information, erection instruction, project management logistics will be managed through a single database, as a collaboration platform, hence the nD BIM model will be continuously used and developed in the Construction & Facility Management (FM) stages (McCuen et al.,

2012, BSI, 2013, Lu and Olofsson, 2014). It has also been argued that the 3D graphical information of the building will have a great impact on the overall BIM capability which brings a focus on the design phases (Giel et al., 2012). As a result, a certain level of BIM business functions overlap during the design stage and connect with later stages e.g. the application of 4th Dimension BIM (4D BIM) in the construction stage and COBie for the operational stages (BSI, 2014b). To facilitate this procedure of work, all stakeholders are required to get involved at the early design stage to clarify their requirements (Mahalingam et al., 2010, Love et al., 2011). It is also believe, BIM performance and its benefits could be increased if BIM can be used as much as possible, as early as possible by all stakeholders during the BIM procedure (Gao, 2011).

The present author believes that by improving the BIM usage performance during the design stage, there could be a positive influence on BIM's performance in the entire project lifecycle. From this, a BIM evaluation framework that is based on industry practices, mainly focusing on the design phase of a project lifecycle is urgently needed as it could have a positive effect to a company's performance (Miettinen and Paavola, 2014). Such an evaluation framework will have the following features:

1. It addresses those changes that are required to transfer from a traditional approach to a BIM based approach (London et al., 2009).
2. It can also be used to apply self-evaluation for the current status: what is the strength, weakness and what is missing for the actual strategy compared to a targeted level. An improved and customised strategy therefore can be developed;
3. Decision Maker could rely on it to make decisions for BIM adoption;

4. The assessment result could be used by the client during tendering and the prequalification stage by looking for the best match service providers.

1.3 Aim and objectives

In summary, the framework introduced in this book intends to improve the use of BIM in the AEC projects mainly during its design stage, as well as in the organizational level. The problem statement is formulated as follow: in order to improve BIM adoption efficiency, an evaluation framework is needed to evaluate adoption status of BIM based on a unified prioritized key implementation area, which also inform a strategy to utilize available resource.

1.4 Outline

Chapter 1 – Introduction: introduces the problem of BIM in the current industry, the need for assessment and the difficulties in doing so. This chapter also knowledge contribution and book structure.

Chapter 2 – Literature review: reviews relevant information to the scope of this book, including: BIM history and current status, BIM implementation frameworks, BIM standards, guidelines and protocols etc. as well as existing BIM assessment frameworks.

Chapter 3 – Framework development: explain the paradigm and rationale of how this framework has been developed.

Chapter 4 – Dimensions & factors for BIM implementation and evaluation: presents the developed BIM implementation Framework (BiF) and BIM evaluation Framework (BeF).

Chapter 5 – Validation and Application: the proposed benchmark has been deployed in real industry projects. Its application, outcome, lesson learned, and take away points will be covered.

Chapter 6 – Conclusion.

2 LITERATURE REVIEW

2.1 BIM introduction

The conventional approach for information management in construction industry could not keep up with expectations for more efficient ways of working due to its limitations and disadvantages (Taylor et al., 2009, Mahalingam et al., 2010). The main aspect missing is the capability of integration of vital information for design evaluation and construction e.g. Bills of Materials, timelines, specifications, price lists, installation and maintenance guide etc. (InfoComm BIM Taskforce, 2011). Moreover, there is a need to share information across all project stages and processes (Rezgui et al., 2013).

The philosophy of BIM was mentioned by Eastman et al. as a 'Building Description System' (Eastman, 1974), and the term 'Building Modelling' was mentioned by Robert Aish in 1986 including 3D modelling, real-time construction simulation (Aish, 1986). 'Building Information Model' was firstly used by Nederveen in 1992 (Nederveen, 1992). The term Building Information Modelling appeared later and was mentioned by Tolman in 1999 (Tolman, 1999).

Initially BIM was defined as a digitalized representation of building and its attributes, a new concept for data, personnel, process and information management (Arayici et al., 2009) during the entire building lifecycle of AEC industry (Eastman et al., 2011, Porwal and Hewage, 2013). It promotes a new relationship and collaboration paradigm among stakeholders. It is also an advanced modelling and simulation concept to improve sustainable design, customer satisfaction and commercial value (Love et al., 2013). In addition, the concept promotes integration, where multiple types of information embedded in the same digital database could benefit and facilitate collaboration among all stakeholders e.g.

designer, contractor, facility manager, etc. (Rezgui et al., 2013). More specifically, BIM integrates the following new functionalities into traditional construction process: project feasibility study, 3D design/drawings, atypical shape design, time line management, costing analysis, clash detection, sustainability analysis, constructability, facility management and engineering analysis etc. (Ding et al., 2014, Lee et al., 2015).

The benefits of BIM have been concluded by academics and construction industry. Some of them have been listed as follows:

1. BIM can greatly improve the project performance and outcome (e.g. sustainability, cost and quality) by solving potential problems at early design stage (Love et al., 2013, Miettinen and Paavola, 2014);
2. BIM transforms the project delivery process for improvement and adds value across the whole project lifecycle (Sebastian, 2011, Wang et al., 2015).

BIM aims to provide project stakeholders (the client, etc.) with sufficient information for their better decision makings. More problems can be solved more easily or prevented entirely at an earlier design stage (Napier et al., 2009).

BIM has already been adopted by countries such as US, Finland, Denmark, Norway etc. and have made considerable progress (Arayici et al., 2009, Smith, 2014). UK released government's strategy in 2011 to introduce the push-pull approach for BIM implementation for all government project with an aim of achieving 20% saving during procurement cost (BIMIWG, 2011, Cabinet Office, 2011). 'Push' aims to improve the construction benefit from the industry side by requiring all users to reach BIM level 2 (fully collaborative 3D BIM, where all project and asset information, documentation and data being electronically managed) (Figure 2-1) by 2016. Relevant 'push' elements include guidelines, training and tools. 'Pull' from the client side aims to improve the post - occupation benefit, where information will be specified, collected and used by the client (Churcher and Richards, 2013). This will require the specific information need to be prepared and delivered to the client on time.

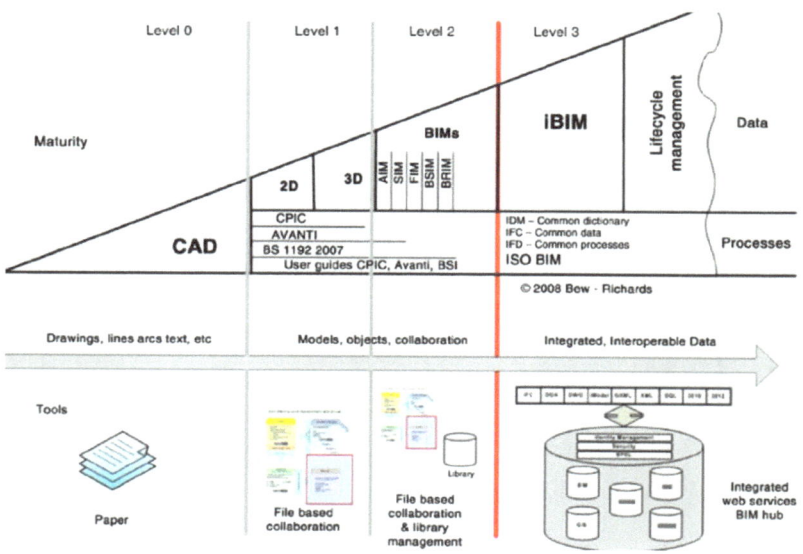

Figure 2-1 BIM maturity levels by Bew –Richards (BIMIWG, 2011)

China BIM Union was established in 2013 as part of the China Industry Technology Innovation Strategic Alliance by the Ministry of Science and Technology. It initiates the development of China's own construction information standards and relevant building information standards (MCIWLG, 2003, MOHURD and AQSIQ, MOHURD, 2014), since it is believed as the key for BIM to be continuously improved in a positive way, projects over 20,000 square meters need to include BIM procedures (MOHURD, 2013). Various standards, guidelines (e.g. Beijing (BMCUP, 2013)) and initiatives for BIM have also been developed in some of the more developed areas in China (e.g. Shanghai (GOSMPG, 2015) and Guangdong (DHURDGP, 2014) etc.). Such standards and initiatives are still at a very basic level, only including IT requirements, level of development (LOD) and data delivery. The official document for BIM strategy of China was released in 2015, with the following requirements imposed by the government: first class design organizations and first class contractors should be able to integrate BIM with their organizational management system and other information technologies by 2020; more than 90% of the future government owned projects regarding design, construction and operation should be BIM compatible by 2020 (MOHURD, 2015).

In summary, BIM covers the whole life of a construction project (Taylor et al., 2009). However, technical aspect development at the design stage brings more tangible profits to users compared to those at later stages. This caused an increased interest in the technical side of things and a reduced one concerning the non-technology aspects of BIM, such as the development of BIM lifecycle (Jung and Gibson, 1999). This therefore leads to a main focus on specific BIM technical applications e.g. visualization, scheduling management, clash detection etc. However, due to the lack of effective guidance, the coordination and management paradigm is still technology driven or even based on traditional approach (Gao et al., 2015). This is not in line with the industry's expectations (NBS, 2014, McGraw Hill Construction, 2010).

2.2 BIM dimensions

With the wide adoption of BIM in the worldwide AEC industry, new areas of BIM implementation and new business opportunities are being explored. The main implementation focus has been shifted from technical only to multi-dimensional aspects e.g. management, process, people, policy etc. (Brynjolfsson, 1993, Jung and Gibson, 1999, Succar, 2009b, Succar and Kassem, 2015). This has greatly expanded the implementation area and also the complexity of the implementation process (Popov et al., 2010).

Technology

Technology has been defined as 'the application of scientific knowledge for practical purposes' (Dixtionaries). BIM can be regarded as an extension of the conventional CAD approach (Singh et al., 2011). As mentioned, a starting point, the focus was laid on the relevant technology improvement which is believed to be the key to speed up the transformation process (Gu and London, 2010). However, such implementation paradigm neglects other non-technological aspects, their roles and functions, which has resulted in a negative effect (e.g. messy talking (Ibrahim, 2013)) and hindered long term development that is required for optimum BIM implementation (Khosrowshahi and Arayici, 2012, NBS, 2014).

Management

The managerial aspects also play a key role as it improves the viability, efficiency and effectiveness of BIM by providing strategic planning, which will also assist decision making as well as prioritize actions (Jung and Gibson, 1999, Tsai et al., 2014). The key business indicators that have been identified by Aranda-Mena et al. (2009) also proved the value of adopting BIM from management perspective.

Process

Process has been defined as 'a specific ordering of work activities across time and place, with a beginning, an end, and clearly identified inputs and outputs: a structure for action' (Davenport, 1992). With the adoption and evolving of BIM, the process of traditional AEC project: its components and relationships between project stages, activities and tasks within each stage, have all been influenced, which include business drivers, automated process analysis and interoperable information that is consistently used throughout the project lifecycle (Succar, 2009a). Recent empirical study even proves that process aspect have a larger proportion than technical aspect (Eadie et al., 2013).

People

People as the instigator and primary medium of revolution, especially their new roles and responsibilities receive equally important attention and function during BIM implementation (Howard and Björk, 2008, Gu and London, 2010). A good leader should be able to inspire the participants of a fragmented and complex project structure, and push towards a common goal for the project, or even for the organization. Hence to improve collaboration, and solve problems among partners (Dossick et al., 2010).

Policy

Policy is described as 'written principles or rules to guide decision-making' (Definition of Policy, 2007). This will include practitioner preparation, problem research & solution, benefit & risk allocation and collaboration among stakeholders etc. Factors contained in this dimension include: contractual and legal standard, regulations and research and development etc. (Succar, 2009a).

2.3 BIM implementation

Lee (2007) proposed four phases of BIM implementation in practice:

1. Phase one: personal adoption — BIM data produced by a single modeller for his own discipline;
2. Phase two: the adoption of BIM in a single discipline within the organization;
3. Phase three: the adoption of BIM in multiple disciplines within the organization;
4. Phase four: the adoption of BIM across organizations and different platforms.

The organization is an integration of individuals (Khosrowshahi and

Arayici, 2012). Since the main challenges of adopting BIM is the organization's readiness to change(Alshawi et al., 2008, Dossick et al., 2010, Khosrowshahi and Arayici, 2012), the willingness from the employee is believed as the main roadblock for BIM using in organization (Dossick et al., 2010, Lee et al., 2015). Moreover, there is also a strong correlation between an individual's BIM skill and the BIM capability of their company (Giel et al., 2012). In order to improve people's willingness to change, their awareness and confidence regarding BIM effectiveness compared to ordinary business activities need to be improved (Singh et al., 2011). Arayici and Coates (2013) emphasize that by focusing on the technical training (e.g. operational skills) of individuals on a daily basis, the benefits of BIM would become more apparent, which will further improve people's confidence in BIM.

Nonetheless, adopting BIM around the traditional process is a 'major change management task' (Khosrowshahi and Arayici, 2012). The main issue for BIM adoption to the next level is the lack of a well-developed strategy to improve the collaboration between disciplines and organizations on an industry wide level, which could maximize the benefit of BIM as well as facilitating innovation (Teicholz, 2013).

Recent efforts have been proposed to improve BIM adoption in the AEC industry from different perspectives. This includes the development of BIM implementation frameworks by researchers, and associated BIM implementation guidelines, standards, execution plans and protocols by government or research institutions.

2.3.1 BIM standards, guidelines, executions plans and protocols etc.

The government is believed as the most effective leading force to push BIM adoption (Cabinet Office, 2011). Meanwhile, national or international research organizations such as NIBS, AIA, buildingSMART,

BSi have investigated methodologies to refine the BIM deployment approach from their own perspective (Arayici et al., 2009, Miettinen and Paavola, 2014). Existing BIM documents (e.g. standards, guidelines and protocols) are available for BIM users to implement BIM into their system (Tsai et al., 2014a). The selection of method according to the user's interest and preference leads to competition and development (Miettinen and Paavola, 2014). The continuous publishing of BIM standards has resulted in *'simultaneously development of standard procedures and tools and their constant reconfiguration locally'* (Suchman, 2007). Such a plethora of BIM documents could potentially lead to diverse methods, maturity and compliance levels of BIM among project partners, which can cause interoperability problems.

A BIM Execution plan e.g. (Penn State University, 2010) reflects the practical implementation of BIM guidelines. It listed those key BIM implementation areas along with who will responsible for each activity, as well as when it will be delivered in an actual project. It intends to achieve an agreement on the plan of the BIM strategy among project partners at the early stage of the project, including BIM users, roles and responsibilities of each individual, deliverables, legal issues, liability and responsibilities and other key points for BIM to be highlighted to all parties throughout the whole project (NIBS, 2015).

By adopting a BIM execution plan, a transparent liability and responsibility among project participants can be established, and therefore reduce potential disputes and improve the collaboration process of the BIM based environment.

In order to align BIM concepts to the contract, BIM guidelines and standards and execution plans can be transferred into a protocol format which will then be endorsed by all partners. Such protocols can provide a general requirement at each stage of work regarding to the deliverables, and ensures all individuals to work in a well-coordinated environment. It also serves as a foundation for BIM to be adopted in a project by providing the structure and support to the company's own guidelines and standards (State of Ohio, 2010).

Regarding the massive amount of documents available in the industry, only a few have been discussed in this book. In order to provide a general idea of what those key implementation areas are or areas that need special attention, the selected documents have been only reviewed generally without detail. For example, one of the guidelines that suitable for both new and renovation projects, the project: BIM Requirements by Senate Properties of Finland (Senate Properties, 2007) identifies the criteria that affect decision making during the modelling purpose of the entire design process, such as design alternative comparison (scopes, costs and energy budget etc.), modelling element explanation of each discipline (e.g. architectural, MEP and structural BIM), quality assurance.

BIM guidelines for different sectors

Other than the above mentioned standards and guidelines for BIM implementation from a lifecycle and all disciplines' perspective, there are also documents that have a focus on a certain discipline's perspective e.g. client and contractor.

Client perspective

Dawood & Iqbal (2010) believes architects are in the position to lead the BIM implementation in order to support the integrated project delivery through control, coordination and management of the project. For the architectural discipline, BIM is more like a digitalized modelling process technology which assists in meeting the client's requirements (TAGCA, 2006, NYCDB, 2013, BSI, 2014a). Regular data drop to the client could keep the client up to date with the latest progress, thus making them aware of design performance, to minimize change order at later stages. Moreover, this could also incrementally improve the client's understanding of the design scheme (BSI, 2014a). The client therefore should be responsible for the additional deliverables and services e.g. as-built models (TAGCA, 2006). Moreover, considering the availability of integrated single database and engineering analysis functions, in a BIM based environment, the client can now procure documentation for performance in energy savings and carbon reduction on top of the usual documents provided by a traditional approach (SECG, 2013).

Contractor perspective

In order to help the contractor to start using BIM, relevant documents have been reviewed (TAGCA, 2006, SECG, 2013): The contractor could adopt BIM from four aspects: tools, process, responsibilities and risk management. The selection of BIM tools should be based on multiple criteria, such as cost, expected functionalities, training and compatibility with the industry trend. Based on the BIM tools, the BIM process should be tailored to produce the expected outcome. BIM could change the way in which people collaborate, share project data and review relevant work. The core of each partners' responsibilities will not be changed but new roles will be allocated to redefine and clarify the new responsibilities among the project teams. Moreover, with regards to the characteristics of BIM: the sharable digitalized collaborative database that to be used and exchanged among all project participants, their ownerships and liabilities are evolving along with the project progress. Hence risk management is another issue need to be considered. For contractor, 'what are the deliverables and who is responsible for them?' are the most important questions.

The review of existing BIM documents (guidelines etc.) has concluded some findings shown as follows:

1. The implementation area for BIM in different disciplines, sectors and project stages have a similar content and coverage;
2. Most information used during construction and later operation stage are based on the BIM database or 2D drawings created in design process, therefore all later applications of BIM are dependent on the design outcome. As a result, clients and contractors are required to participate in an earlier design stage to clarify their demand of what kind of information is to be embedded within the design model.

However, some disadvantages of surveyed documents (guidelines etc.) have been identified in practical usage:

1. Existing guidelines etc. have been developed based on the local workflow, supply chain readiness, infrastructure etc. which may not be suitable to other regions;
2. Key implementation areas have merely proposed, without thorough priority or adoption sequences according to a specific project's requirements; they are sometimes based on intuition. This type of adoption could lead to budget overrun loss of focus. What's more, a more effective strategy plan to meet business objectives is needed (Tsai et al., 2014a);
3. Different sectors/disciplines will have their own focuses, this could lead to a different prioritising sequence in term of BIM adoption;
4. The project outcomes, limitations and drawbacks of the implementation process for using existing guidelines cannot be evaluated (Kam et al., 2013a, Tsai et al., 2014a).

2.3.2 BIM Implementation Framework

A framework represents a holistic relationship of terms and concepts of a system (Jung and Gibson, 1999). It aims to sort out the logic structure of the interested system as well as the future development direction (Jung and Joo, 2011). Similar to the framework, the model can be defined as the representation of a reality in a specific environment (Dehe and Bamford, 2015). By developing such a framework for BIM, it will facilitate all practitioners to be fully aware of the composition of the BIM concept. By considering BIM as the integration of final product and the process of delivering it, users will begin to understand, disseminate and incrementally implement BIM. In addition, a framework that could achieve 'presenting data and arguments in manageable sections' could fulfil the gap remains between academic and industry knowledge bodies (Succar, 2009a). A number of BIM collaboration frameworks have been briefly introduced in the following sections.

Information System (IS) is one predecessor that is similar to the concept of BIM, which is believed to improve the effectiveness of AEC projects. However, the concept of IS lacks a clear strategy with focus and objective. The strategy and IS could complement each other: the strategy could serve as a clear future direction and development guidelines for IS, while IS could explore more and more business opportunities and market which influence the organization's strategy. CIC calls for the integration between IS and the AEC industry to improve its productivity. However, this undertaking emphasizes only the technical aspects. The industry and practitioners therefore lack knowledge on how to implement CIC, to what extent it should be developed and what is the prioritizing sequence of demands (Jung and Gibson, 1999).

Therefore Jung and Gibson (1999) proposed a planning methodology for CIC implementation from company level. The CIC framework includes three dimensions: Information System (IS) Concern (4 areas of concern), Business function (14 functions) and project lifecycle phases (6 phases). In order to provide an objective and quantitative assessment result, five main areas for the assessment have been proposed: corporate strategy, management, computer system, IT and incremental assessment, these five areas include both business and social-organizational issue. Moreover, business functions have been considered as evaluation criteria for effective CIC planning. Analytical Hierarchy Process (AHP) was proposed by the authors to assign relevant weight to those measures in future work. The individual assessment of these five main areas could effectively assist decision making during implementation in their own domain, and together as a whole. Instead of making a decision, they could also balance resource investment on an organizational level, identify their departure point and objectives. However, in this research, Jung et al. have not considered cross organization collaboration or data management issues. Based on the connection between BIM and CIC, as a follow-up study of (Jung and Gibson, 1999), Jung and Joo (2011) proposed a framework to systematically represent relevant areas of BIM practical implementation. The thesis author argues, before the development of an assessment framework, these assessment criteria should be validated for their applicability by industry practitioners. The proposed framework is not comprehensive enough either. One limitation for example is that degree of involvement of all partners during BIM implementation should be considered. Moreover, the framework requires an update under the 'function' category as follows: business function during the operation stage should be considered, since the information required for building operation purpose should be prepared in early design stages. Most importantly, there is no follow up research to develop an evaluation methodology based on Jung's framework.

The approach mentioned above has been summarized as top-down pattern for CIC implementation. Top-down pattern intends to curtail the adoption of CIC by partially select adoption scope according to the priority of managerial prioritize and organization's strategy. This will have a more effective outcome compared to a full-range adoption. Based on the identified limitations of CIC and IS, the key of BIM success is to have a tangible and practical BIM strategy and prioritized key implementation criteria across all aspects (PSCIC, 2013).

The top-down approach could also be aligned with the 'touch the BIM lightly' approach: a relatively more progressive and acceptable paradigm for technology proficiency, which resulting in a more focused and effective learning process (Ibrahim, 2013, Forsythe, 2014). Hartmann et al. (2012) also agrees that a top-down approach could effectively integrate BIM with the organization's strategy and could strategically plan for BIM as a goal for large scale and long term implementation (Arayici et al., 2011). Dikmen et al. (2005) revealed that organization strategies are one of the main driving forces for achieving organizational effectiveness.

Khosrowshahi et al. (2012) believed that to implement a new concept like BIM, it is important to identify the core of the problem, the current status of existing approaches as well as an example to follow. The culture, personnel and technologies of organizations are different, which require a specific strategy for a better implementation. The author conducted a mixed of qualitative and quantitative method to identify the industry's readiness. First, a literature review was employed to reveal the current practice of BIM, include both driving forces and barriers of adoption. Secondly, a BIM implementation concept map was developed based on the interview with BIM users in Finland. Khosrowshahi et al. argues this map is believed to be a reference object for UK BIM implementation. It includes three focus points: organizational culture, education and training and information management. Thirdly, a questionnaire approach was adopted in UK AEC industry to identify people's perception and level of acceptance for BIM in UK's AEC industry. The questionnaire's result could derive another concept map specific for UK's status. Lastly, a road map based on the results from the three research steps has been developed: provide a feasible implementation strategy and guidelines as well as find answer for breaching the barriers and challenges which came to light during the questionnaire.

There are some defects of the above mentioned research. The roadmap summarized from interview session contains a number of criteria. The importance level of these criteria have not been assessed by all interviewees to check their consensus, since some of these factors might disagree with others. Khosrowshahi and Arayici (2013) claim the final road map which provide a list of 'implementation areas' could allocate limited resources in a better way. However, without prioritizing this list, it will be hard to decide what is more important and what needs to be done first. This could lead to waste of resources, or even confusion on where to start, hence the effect will be limited. Moreover, the author fail to assess if what have been summarized from Finland is adopted in UK due to the differences between these two countries, such as culture, policy, infrastructure etc.

2.4 BIM Assessment and Benchmarking

2.4.1 BIM Evaluation & assessment

Kwak and Ibbs (2002) proposed an idea to assess project management, they define a good project management (PM) as 'a well-defined level of sophistication that assess an organization's current project management practices and processes'. They also believed that the composition of assessment approach used to assess organization's current practice has become sophisticated over the years. It therefore leads to confusion, uncertainty and difficulties in assessing their current practices. Hence a $(PM)^2$ model was developed as a reference point and yardstick to identify the current status and weakness of PM practice within the organization. In addition, it also acts as a systematic and incremental approach, which guides project managers to transit from unsophisticated to sophisticated levels.

The development of assessment method in BIM is lagging behind than other domain in the AEC industry (Kam et al., 2013a). For example, for green building evaluation/assessment, there are a number of environment assessment framework available on the market, for example, BREEAM, LEED etc. These methods are generally divided into three parts: the first part introduces the goal of the assessment, followed by the second part, categories of concern; the last part deals with the assessment criteria in each of the described categories (Chang et al., 2007b).

I-CMM BIM assessment framework:

As the earliest and most used assessment framework in the US, the National Institute of Building Science proposed the BIM Interactive Capability Maturity Model (I-CMM) based on 11 criteria (data richness and information accuracy etc.) with 10 capability maturity levels for each. It intends for 'users to evaluate their business practices along a continuum or spectrum of desired technical level functionality' as well as 'for use in measuring the degree to which a building information model implements a mature BIM Standard'. Regarding its single aspect of assessment in information management, it is not for any benchmarking purpose or for 'BIM implementations comparison' (NIBS, 2007b, Kam et al., 2013a).

The evaluation result can be affected by many variables which could potentially reduce the effectiveness of the assessment method (Succar, 2009b). Firstly, the weighting scheme for the maturity level of each criteria can be modified freely by the user (NIBS, 2007b, Succar, 2009b). However, there is no scientific calculation method to provide the most reasonable weighting system; as a fact, the weight was obtained by vote (McCuen et al., 2012). Secondly, the lowest credit required to meet the 'minimum BIM' will change with time: 'as the rhetorical bar is raised and owners demand more from the models being delivered' (NIBS, 2007b). On the other hand, the accumulative credit gained from the I-CMM is not reliable, since a higher result does not guarantee every aspect of the evaluation could achieve the associated maturity level. There are ten maturity levels for each of those evaluation criteria, while normal assessment methods will have a maximum of five to six maturity levels (Succar, 2009b). This could confuse the assessor during maturity level selection. These ten criteria have not been explained well for practical usage, and there is also a definition overlap, in some cases (Suermann et al., 2008).

The I-CMM has been adopted to assess the AIA-TAP BIM Award winners in both 2007 and 2008 (Suermann et al., 2008, McCuen et al., 2012). However, only visualization aspect has been tested, while the accuracy of other areas of BIM adoption have not been tested.

BIM proficiency Matrix (BPM):

In order to evaluate the individual's BIM skill proficiency, for both designers and contractors (Kam et al., 2013a, Giel, 2014), Indiana University developed a BIM proficiency Matrix (BPM) with eight categories and each category has been divided into four maturity levels (Indiana University, 2009b). A score is also presented with associated certifications. From the present author's view, there is not enough information available for research purposes or validation processes to test its validity. Kam believe this assessment method lacks social aspect consideration (Kam et al., 2013a).

BIM3 – Succar BIM assessment framework:

Succar (2009a) divides the BIM field into three parts: technology, process and policy; each dimension includes both players and deliverables. A push-pull relationship will be formed between any two parts so as to deliver the required information or business activities e.g. contractual relationships and project deliverables. Succar also defines BIM stages into four, starting from non - BIM to the current ultimate BIM level: integrated BIM. It expresses a gradually and continuously implementation maturity of BIM stages in term of process, technology and policy.

Based on this framework, Succar (2009b) developed a BIM Maturity Matrix (BIm^3) as 'a knowledge tool which incorporates many BIM Framework components for the purpose of performance measurement and improvement'. BIm^3 contains five components (Succar et al., 2012):

1. BIM capability stages: 'is defined here as the basic ability to perform a task or deliver a BIM service/product';

2. BIM maturity index: 'to reflect the specifics of BIM capability, implementation requirements, performance targets and quality management.'

3. BIM competency sets: 'reflects a generic set of abilities suitable for implementing as well as assessing BIM capability and/or maturity.' In more detail, it can be classified into three groups: technology, process and policy to align them with the BIM field (Succar, 2009a);

4. Organizational hierarchy and scale, 'To allow BIM performance assessments to respect the diversity of markets, disciplines and company sizes';

5. BIM granularity levels to increase 'assessment breadth, scoring detail, formality and assessor specialization' along with the increase in granularity. In total, they have included area such as: software, hardware, products & services, contractual and organizations etc. (granularity level 1).

However, there are still defects to be addressed further by the present author.

Firstly, only three dimensions: policy, process and technology are included. Each dimension has stakeholders, such as regulatory bodies in policy dimension, owners and designers in process field and software companies in the technology dimension. With the consideration of multiple dimensions of BIM, as well as the equally important levels of these dimensions, there is a need to re-categorize some factors within these three dimension (e.g. people) into separate dimensions (e.g. people dimension), to form a clearer pull-push relationship under BIM interactions and BIM field overlaps.

Secondly, the assessor needs to select the scale of their organization as well as their benchmarked granularity level 'to enhance BIM capability and maturity assessments process and to increase their flexibility' (Succar et al., 2012). Compared to less granularity, the selection of higher granularity will have more assessment criteria to be used. Succar argues this could be suitable for various purposes and different kinds of assessments. Informal/self-assessment will prefer a more abstract level of assessment, while 'specialist-led appraisals' could prefer a more detailed assessment.

Thirdly, the number of 'competency areas' can also be influenced and varied by the organizational scale and capability stage (Succar, 2009b). However, the author of this thesis argues this type of 'flexibility' might cause inconvenience and inaccuracies during the assessment as the actual 'competency area' of BIM is unpredictable in actual practice. For example, an organization's major BIM 'competency area' is in stage 1, however, they may have met the standards of some assessment criteria that belonged to higher granularity level, such as stage 2 or even stage 3. In this case, the selection of stage 1 could not provide an accurate result.

Lastly, Succar only published those 'competency area' for granularity level 2 (Succar, 2009b) and 'competency area' for technology dimension for level 4 (Succar et al., 2012). Moreover, the BIM maturity matrix is only based on granularity level 1 (Succar, 2009b), hence there is a lack of comprehensive explanation for overall assessment criteria. Sebastian and Berlo (2010) comment that once the minimum requirement of capability in a granularity level has been reached, there will be no further comparison until the next stage of maturity matrix has been used and achieved. Sebastian and Berlo also argue the proposed framework could not compare the experience and modelling quality of BIM in different organizations, even they are at the same stage. Kam et al. (2013a) also comments there lacks of validation of the proposed framework.

BIM Characterization Framework (Gao, 2011):

Gao (2011) proposed a characterization framework for BIM, with the intention to understand how BIM should be conducted and who should be involved. This framework has divided BIM based project information into 3 categories, 14 factors and 74 measures.

Gao argues the adoption of BIM in a single case study is inconsistency and insufficiency. It requires data to be collected from large amount of case studies. Data from 40 case studies have been collected and analyzed to finalize a comprehensive list of criteria which can be used as benchmarking by other users to compare their work with best practices. New 'implementation criteria' could be identified and added into existing framework. It could also be repetitively used by the same organization, with experience accumulated, to reveal the most often used criteria as top priority; a template of BIM adoption can be developed for future use.

According to the present author, this research also has its limitations though: (1). The criteria within the framework needs to be updated e.g. 5D costing control has not been considered; (2). The framework is only applicable for BIM implementation process on project level, not organizational level; (3). It targets at design and construction stage, while the information required by operational stage should be considered as well in early design stage; (4). The collected user perceptions on BIM performance are not for evaluation purpose. (5). The proposed framework has only been tested using already completed projects. Its value and benefit on new projects has not been tested.

Organizational BIM assessment framework (Kreider 2011):

With the emphasis on the top-down BIM implementation paradigm from the management level (Jung and Joo, 2011), the use of BIM within the organizational level received a higher emphasis from BIM users. A BIM maturity framework from client/facility owner's perspective was developed by Kreider (2011) for organizational BIM (OBIM) usage. This assessment framework contains six main areas: strategy, uses, process, information, infrastructure and personnel's BIM competency. This assessment method is believed as effective (PSCIC, 2013). However, the present author believes the result provided could be inaccurate, as it assigns equal weights to each maturity level and each criteria, and that will not be able to identify the priority of assessment criteria. Besides, the method also lacks relevant assessment in data management, BIM application and stakeholders' involvement. Finally, how those criteria have been selected and whether they are empirically validated or not is unknown.

ABMF Maturity framework from Arup

Andrew Duncan and Graham Aldwinckle from Arup proposed their own BIM project Maturity framework (ABMF) to assess project management aspect within four primary disciplines: SMEP (Structural, Mechanical, Electrical and Public health) and 21 secondary disciplines (Acoustics and Fire etc.) (Duncan and Aldwinckle, 2015). There are 11 questions (Common Data Environment and reference/version control etc.) and 3-6 maturity levels for assessors to select for project management, and 12 criteria (e.g. LOD, 4D and 5D etc.) for other primary and secondary disciplines. In these 12 evaluation criteria, one criterion is discipline specific, the rest remain the same.

The ABME framework was mainly designed to assess BIM process based on the work done by Penn State University. Those assessment criteria are in majority for design related disciplines, which might not be applicable to other secondary disciplines e.g. public health.

Arup intends to use ABMF to compare their own worldwide BIM projects. According to the present author, the selected criteria and weight can be influenced by the local cultural and government policy in company level. While there is a lack of relevant information on how those criteria have been created or selected, their applicability towards users from different regions has not been described, and no empirical evidence could show the reliability of the assessment result. Additionally, how the weight of each criteria was obtained has not been explained.

2.4.2 BIM Project Benchmarking

Benchmarking is a structured method to measure and compare an organization's processes, activities, and performance against others organizations (Garvin, 1993). By comparing with other peers' best practice, it could reveal the defects of current practice and an appropriate improvement strategy can be planned and proposed accordingly (Camp, 1995, Succar and Kassem, 2015). The benchmarking process will gather a group of decision makers from different organizations, who will share knowledge and experience amongst themselves to encourage innovation (Garvin, 1993, Costa et al., 2006). The result achieved can be used to rank the organizations' competitive position within the industry level and the organizations' possibility of success (Tsai et al., 2014a).

However, most existing assessment framework are only suitable for evaluation but not for benchmarking, and there is no available national or international BIM benchmarking instrument existing yet (Sebastian and van Berlo, 2010, Du et al., 2014).

Sebastian and Berlo (2010) proposed a BIM benchmarking tool including 4 main perspectives: organization and management, mentality and culture, information structure and information flow, tools and applications. These perspectives have been further divided into following factors: vision, roles and organization structure etc. Interviews and desk research have been used to formalize the framework. According to the present author, there are some limitations for that framework:

1. During the first research step, desk research might intake some criteria which haven't been reviewed during the interview session, which need further validation;
2. The use of this benchmarking tool requires certified professional users, which will hinder the dissemination industry wide;
3. As a benchmarking tool, Sebastian and Berlo fail to address how the selection of assessment criteria and their weightage calculation could meet the requirements, visions, strategies and implementation focus of different organizations;
4. The accuracy of the proposed tool has not been proved. The validation process includes two organizations, the author only mention 80% of assessment criteria received a consistent result for the fictional organization, but do not mention any result for the real organization.

CIFE developed Virtual Design and Construction (VDC) maturity evaluation scheme focusing on four areas: planning, adoption, technology and performance, which aims to provide quantifiable and qualitative measures. It uses scorecard to evaluate its market, performance, barriers and future development direction. The method can also be used for BM assessment (Kam et al., 2013a).

The assessment framework has been further divided into 10 subdivisions (e.g. objective, organization and process etc.) and 56 measure items. In order to enhance the reliability of the result, confidence level with seven factors have been adopted to assess the accuracy of the information of an assessed project: multiple stakeholder input, timing & phase of engagement etc. In order to validate the proposed framework, 108 projects from all over the world has been applied to test its applicability. The result shows the scorecard improved those projects in term of cost saving, shorter schedule, quality and communication, however, there is no solid evidence to prove this improvement. Moreover, the scorecard will continuously evolve along with the technology (Kam et al., 2013b). Regarding the limitations of this research, the present authors believes:

1. The selection of evaluation criteria has not been validated. The percentage assigned to each area and division lacks explanation. There is no weight has been assigned to each criteria. There is also no standard to judge if the confidence level has been met;
2. The proposed Scorecard is not convenient or reliable for use: the evaluation requires qualified people to interview multiple managers from the organization, which is subjective. It will also be difficult to decide the true result out of collected results (Du et al., 2014);
3. It will be used in a relatively later stage which is near project completion, while no guidance or priority can be given at the beginning of the project;
4. Kam et al. also realised the evaluator might not be able to complete the evaluation process, as it takes too long.

A cloud-based BIM performance benchmarking application: BIM cloud score (BIMCS) was proposed by Du et al. (2014). This model is based on software as a service (SaaS) to facilitate the data collection and analysis in a dynamic and automatic way. It can be used as an add-in for design platforms (e.g. Autodesk Revit). The model includes six categories: modelling productivity, effectiveness, model quality, accuracy, usefulness and economy.

The first two aspects collect the performance of the product (BIM model) and the rest collect the performance of the process (BIM process). People's daily operation will be monitored and recorded, with a data mining technique, a large amount of information can be collected in relation to BIM performance, BIM application areas and marketing sector. The evaluation criteria was finalized based on literature review and survey. The weighting system for each criteria was first decided through the use of a survey (1-5 scale rating). The weighting system increases in accuracy as the number of users in the aforementioned survey is increased.

According to the present author, the limitations of the research (BIMCS) are:

1. During the criteria developing stage, the removal or addition of criteria was based on the respondents' own judgement; the changes have not been reviewed with others. Moreover, the initial subjective weight obtained from the five-point Likert scale is not accurate;
2. Evaluation criteria is focusing on the modelling perspective, and there is a lack of process, administration and organizational perspective;
3. The performance of the process or product is based on increasing (e.g. building objective in 3D BIM model) or

decreasing value (e.g. warnings for error) of the 3D BIM modelling process. However, this might be influenced due to external variables e.g. the complexity of the project, change order from the client etc.
4. The data collection approach has decided the benchmarking object is limited to the technical aspect of n-Dimensional BIM model development process and completed products.

2.5 Summary

The review of existing BIM implementation frameworks, standards, guidelines, execution plans, protocols etc. justified that more effective BIM assessment / validation / benchmarking frameworks are needed. The existing BIM assessment frameworks cannot address the issues concluded above; according to the present author, their limitations have been concluded as follows:

1. The source of assessment criteria is mostly based on theory, such as literature review. In order to improve their reliability in practice, questionnaires have been used to obtain empirical evidence (Won et al., 2013). However, participants' suggestion and consensus should be considered for the most comprehensive implementation list (Chen et al., 2014);
2. The weightage of each criteria is based on an estimation value instead of scientific equation (Jung and Joo, 2011, Becerik-Gerber et al., 2012). Their impact towards company's objectives have not been considered to provide a more efficient strategic planning approach;
3. Existing frameworks normally emphasize a single or several aspects of BIM lifecycle adoption (Hartmann et al., 2012, Love et al., 2013, Tsai et al., 2014a). A multi-level organizational approach should be applied to identify the changes to an organization caused by BIM. Existing methods

cannot complement each other due to the assessment criteria and weights are incompatible. Without covering all the relevant implementation issues (Hartmann et al., 2012, Love et al., 2013), it will not be appropriate for new BIM users (Shuo and Jiancheng, 2014);
4. The evaluation framework should be easily updated since its boundaries are evolving in two aspects: software solutions expansion and organizational information systems upgrade (NIBS, 2012, Kam et al., 2013a, Love et al., 2013);
5. Some of the existing methods have not been validated by practical adoption.
6. The relation between each criteria and how they can influence a project's focus scheme are still unclear. By becoming aware this, the decision maker could have the most optimized implementation strategy for improving efficiency (Tsai et al., 2014a);
7. Existing benchmarking tools that been used for cross organization comparison are not perfect, since most methods' weights were firstly obtained based on a questionnaire, and recalibrated by results obtained from future projects. Considering different strategies could influence the implementation priority of different organizations, the weightage achieved in this way cannot provide a fair comparison environment;

Some of the assessment methods and benchmarking tools are not free, which can potentially be ignored by a large amount of applications and reviewer, affecting future improvements.

3 FRAMEWORK DEVELOPMENT

The proposed framework was developed based on pragmatism philosophy, supported by a mix approach (quantitative and qualitative research) where social and other contexts have been comprehensively considered within the research environment (Wicks and Freeman, 1998, Meesapawong, 2013).

3.1 Framework Design

In order to provide a consistent and accurate assessment method to make a good decision on the investment in BIM, to improve BIM performance and project outcome, a multiple criteria decision making (MCDM) framework is needed. The MCDM method provides a framework to handle both quantitative and qualitative objectives (Sonmez et al., 2002). This framework can be treated as a platform where all stakeholders could exchange their points of view by following four steps: structure the problem, priorities the criteria, evaluate and make final decisions (Bozbura et al., 2007). The MCDM has been applied in project management, pre-qualification (Sonmez et al., 2002) and organizational self-assessment (Yang, 2001, Xu and Yang, 2003).

In this book, the author argues BIM implementation in project and organization management is a process of decision making among a complex integrated environment, which involves both quantitative and qualitative information. MCDM relies upon the constructivist knowledge, which is the interaction between the object (research problem: BIM implementation & assessment) and the subjects (decision maker, DM) (de Moraes et al., 2010).

Research that has a close connection with a real-life context can be validated by empirical evidence through case study methods (Oates, 2006). By adopting a real case scenario, the research assumption can be tested in experiments with sufficient information to be collected (Won et al., 2013).

BIM is a highly practical concept, it should therefore be considered equally important to analyze BIM theoretically (Chen and Li, 2014). The proposed BiF for BIM implementation in the previous step needs to be applicable to industry practice (Kam et al., 2013a). After such a MCDM framework was developed, names as: BIM implementation Framework (BiF) that includes five dimensions and sixty nine criteria based on existing BIM documentations from literature review. In order to test their applicability in the industry, the proposed BiF will be tested on a real project, to collect industry perception, awareness and readiness of BIM.

3.2 Research context

The implementation of BIM in AEC projects will consist of a huge amount of interaction among stakeholders, project stages, activities under certain constraints e.g. time, budget and sustainability requirement etc. Other than the transformation of technical dimensions, social aspect, such as social-organizational culture will also be involved (WSP, 2011, Khosrowshahi and Arayici, 2012). As such, there is a need to improve BIM practical adoption from a social perspective to provide practical ways of managing BIM in a specific context.

It is estimated that in 2020, China will carry out 23% of the world's construction projects (Crosthwaite, 2012). With the consideration of low productivity (Adriaanse et al., 2010), a vast number of the resources consumed have been wasted (The Climate Group, 2014). Based on the low maturity of China's AEC industry (HIG, 2015) and the demand for sustainability in the rest of world China is lagging behind (Wong and Kuan, 2014). The huge amount of construction in China could lead to a 'real momentum towards innovation', where BIM is believed as the leading role (CIOB, 2015). However, China's BIM application is still below acceptable levels (Smith, 2014, Miller and Luo, 2015) which is predominantly due to the lack of a systematic management approach on BIM implementation (Gao et al., 2015) and organization's readiness (Alshawi et al., 2008, Dossick et al., 2010, Khosrowshahi and Arayici, 2012). Therefore BIM is urgently needed in China's AEC industry. Therefore a full scale case study was conducted in China. Hence China has been selected as the context for the research development and validation.

3.3 Framework refinement

It is also necessary to refine these criteria with consideration of industry users' feedback to ensure no missing element remains behind, a consensus based approach: the Delphi method (a multiple rounds of questionnaires approach) was selected to further refine the preliminary framework in a specific context. Delphi consultation was originally developed in the 1950s by the RANO Corporation in one of its projects done with the US defense construction and aimed to 'obtain the most reliable consensus of opinion of a group of experts ... by a series of intensive questionnaires interspersed with controlled opinion feedback' (Linstone and Turoff, 1975).

The initial adoption of Delphi aims to forecast possible varieties in future (Bender et al., 1969). Bender et al. argues that the present is a summary of the past, therefore, by summarizing the present, it could effectively predict the future: Delphi is also believed not only to explore short term decision making but also provide an accurate result for a long term prediction (Ono and Wedemeyer, 1994). Hence it aims 'to determine, predict and explore group attitudes, needs and priorities' (Hasson and Keeney, 2011).

In this research, classical Delphi has been selected: as the most fundamental Delphi method aims to collect participants' judgment, analyze it and to be used as a reference in the next round. Normally this process will continue for a minimum of three rounds until consensus among all participants has been achieved. In terms of the communication approach between the researcher and the participants, postal services was be adopted. This type of Delphi meet the demand of this research: to keep the entire process confidential, and participants could provide the most correct and honest answer without hesitation.

In order to assess the applicability of the proposed BIM implementation Framework (BiF) in a specific context, especially as the initial developed framework may lack latent variables which are difficult to identify or where criteria may overlap with each other, an empirical method is needed to collect opinions from a wide range of practitioners from all disciplines (Du et al., 2014).

The BIM implementation in the AEC industry should adopt multi-dimensional decision making (e.g. project management, organizational management etc.) and stakeholder involvement (e.g. client, designer and contractor etc.). It is important to consider all individual opinions regarding the selection of implementation criteria for decision making purposes. The finalized criteria can therefore be used by a single decision maker instead of a large number of individuals (Sebastian and van Berlo, 2010, AIA, 2013b, Jeong et al., 2013).

Delphi is suitable for multi-dimensional and MCDM problems; it has been adopted for AEC domain to support assessment criteria selection. Hence in this research, the Delphi method were chosen to refine the developed preliminary BIM implementation Framework (BiF).

3.4 Priority and sensitivity analysis

The implementation of BIM within AEC projects involves activities from multiple phases and stakeholders from several disciplines. The activities and associated processes are complex, fragmented and the assessment and evaluation of BIM requires multiple dimensional consideration (Taylor and Levitt, 2007). BIM is at an early development stage which can be defined as "New Frontier Technology", with the following features: technical uncertainties, market risks, lack of hard data, subjectively and intuitively of evaluation process due to the lack of adequate evaluation criteria and a qualified evaluator (Hsu et al., 2003) and a list of accepted criteria which could facilitate BIM implementation and assessment (Zhu and Xu, 2014). All these have made the evaluation and decision making process for BIM industry implementation very complex.

The development of an assessment method requires two steps (Shapira and Simcha, 2009, Won et al., 2013, Alyami, 2015). (1) Step 1 involves the selection of assessment criteria based on a reliable method, followed by the development of a weighting system. This has been addressed through Delphi method as elaborated in previous section.

(2) Step 2 involves a reliable, dependable and applicable weighting system which should be developed to represent the interest of all decision makers from all disciplines and organization levels (manager and technical engineer). An analytical hierarchy process approach was initially introduced by Thomas L. Saaty, as a multiple criteria decision making method (SAATY, 1987). AHP breaks down large and complex problems into small sized factors which are easy to control and manage. By demonstrating these factors into a multi-level hierarchy model, a visual diagram can be created, which will provide the user with a better understanding of the inter-relationship of the entire assessment framework. After the hierarchy model was established, the weight of each criteria will be calculated to demonstrate the relationship amongst all criteria (Wang et al., 2005, Lee et al., 2013).

The hierarchy framework was also designed with a few potential options.

Multi-criteria decision making (MCDM) has been widely used to address engineering, business management, science and technology development issues (Mardani et al., 2015); while AHP as a tool to realize MCDM has been widely applied in AEC industry from both managerial and practical aspects (Subramanian and Ramanathan, 2012, Dehe and Bamford, 2015). AHP can be used for alternative selections for multi-dimensional problems (Jeong et al., 2013, Lee et al., 2013) and assessment purposes (Shapira and Simcha, 2009, Shapira and Lyachin, 2009, Hijazi et al., 2009).

The AHP method could also have the capability to consider the variety of objectives/focus schemes of the organization. The ranking order of each focus scheme can be obtained to highlight the user's current attitude. In addition, the use of AHP application is rising in developing countries for the evaluation of complex systems (Vaidya and Kumar, 2006). Based on the features of the AHP method, it was adopted in this research step.

The MCDM problem is normally an evolving process (Triantaphyllou, 1997). The requirement and demands of the AEC marketplace varies with time, technology and social needs, the organization's focus will change and alter accordingly (Zhu et al., 2005). In a traditional approach, decisions and strategies are developed based on the most important factors, however, through some decision making methods (e.g. AHP), a more subtle case can be detected, where less important factors may be considered as insignificant to the expected target (Subramanian and Ramanathan, 2012).

Sensitivity analysis has been widely used to assist decision making in the AEC industry: a series of well-structured and prioritized criteria is available and their influence effect towards the company's various objectives are provided, understand how each objectives' ranking can be affected due to the shifting of priorities of criteria (Triantaphyllou, 1997). Decisions made under such circumstances could provide a more efficient strategy that continuously add value and will pass this down the management chain (Barry et al., 2012, Subramanian and Ramanathan, 2012, Jeong et al., 2013).

In this research, the priority of BIM implementation areas has been changed to meet the organization's new focus scheme in order to improve the emphasis on sustainability and customer satisfaction. Therefore, sensitivity analysis (Dantzig, 1963) has been conducted in this research, a group-based decision making method known as the Analytical Hierarchy Process (AHP) method was adopted here as a sensitivity analysis can be carried out to develop new strategic plan (Subramanian and Ramanathan, 2012).

Each AEC project is unique (Wegelius-Lehtonen, 2001). This uniqueness is caused by variables such as project location, types of project, contract type and client character etc. (Succar, 2009b), which could all lead to differences. However, there are also similarities led by the unified process of conducting a project, unchanging organization structure, risk measurement, andante knowledge, and experience (Succar, 2009b).

One of the top ranked International Design Consultant – ASZ (its name is anonymous) was selected as the study object. ASZ has more than 35 years of planning, designing, engineering and consulting experience in China, it is operated by local people and working with local partners (clients, contractors and government agencies etc.) which makes ASZ very familiar with the culture, policy and business infrastructure of China. The selected company for the AHP study needs to have a certain level of BIM knowledge and experience in managing BIM based projects. BIM concepts should also have been adopted within the entire organization, even at an initial stage, therefore to be able to propose practical and reasonable BIM project objectives.

With the shrinking of global economics, ASZ currently aims to improve their emphasis on new focused schemes: customer satisfaction and sustainability through comprehensive adoption of BIM, while the transformation has not been successful due to the lack of information or strategy on how to start. Moreover, their decision for a BIM strategy in their current BIM based projects are still based on intuition and without a proper and coherent answer (Jeong et al., 2013). Considering the multiple focus schemes and limited resources, ASZ needs a decision making framework to support a more effective BIM development plan. Therefore, the AHP method was conducted to reveal how each criterion of BIM implementation could influence various focus schemes.

3.5 Framework validation and application

The aim of this validation step was to prove if the proposed BIM evaluation Framework (BeF) could accurately reflect the level of BIM usage in a specific organization and to test its efficacy (ISO Standard, 2010) and user satisfaction, needs and future development (Gupta, 2013). It also intend to reveal the relationship between the use of BIM and project performance: whether a higher level of BIM adoption could have a positive impact on the project performance or not. Moreover, external variables that could influence assessment results also need to be identified.

Three rounds of validation questionnaire were produced by working with BIM manager from ASZ. In round 1, vertical comparison was developed to compare different projects' performance by the same team completed during different periods; horizontal comparison was also produced to compare several parallel projects completed by different teams in similar periods. Five projects have been selected from ASZ.

Validation round 1 questionnaire design

The validation round 1 is based on the questionnaire to evaluate each project based on the following three phases:

1. Phase one. To evaluate BIM usage based on the proposed BeF, the user (e.g. BIM manager) would simply select levels that best meet each project;
2. Phase two: ASZ management team is also interested to know the quality, comprehensiveness and completeness of the nD BIM model that been created and used through the design stage. Since this could reflect user's knowledge, awareness and capability on the development, maintenance, application and management of BIM model, as well as whether those criteria included in the BeF have been truly reflected in the actual result of real project (Sebastian and van Berlo, 2010).

    In this second phase, questions were asked regarding the nD model's naming system, Level of development (LOD) etc. (please refer to Chapter 6 for more detail). User (e.g. BIM manager) would use $1 - 10$ Likert scale, where 1 presents the adoption of this particular aspect of nD model is extremely poor, and 10 presents an extremely satisfied level.

The judgment in phase one & two is ideally to be objectively judged. However, the result still can be influenced by many variables.

3. Therefore, the evaluation of project's performance based on various Key Performance Indicators (KPIs) will be included in phase three.

The comparison of those results obtained by all phases could effectively reveal the relationship between BIM levels and project's performance. In order to improve the initial evaluation package to be more reasonable and more comprehensive, feedback was returned to the authors from the BIM manager in ASZ. After appropriate improvement, the final validation package has been send to the evaluator.

Validation round 1 questionnaire design

In round 2, any missing / additional information (e.g. project information) from round 1 was identified. Existing assessment tools (e.g. I-CMM, BPM, OBIM & ABMF) were used to assess all selected projects in round 1. Their results were compared to the result obtained by using the developed BeF in round 1. In round 3, user feedback was collected concerning the proposed method. In addition, the result of sensitivity analysis was discussed with ASZ for a new BIM strategic planning to meet their new focus scheme.

Evaluator selection

The user based evaluation is crucial as the result and feedback help to improve the final product to be more effective, efficiency and satisfied product. There is no specific requirement for the number of evaluator (Gupta, 2013). However, there is a difference on the evaluator's expertise and position. The developed BIM implementation and assessment framework is intended to be used by BIM management personnel, hence the evaluator used in this research is ASZ BIM manager since he is the only person who can provide the most objective assessment result.

3.6 Summary

This chapter comprehensively explained the generic research methodology and specific methods used in the research conducted by the author. An overall research methodological framework including six steps has been devised to address the concluded research questions. The research follows a pragmatism paradigm with a mix of a qualitative and quantitative method. The six research steps includes one theoretical study and five empirical studies.

4 DIMENSIONS & FACTORS FOR BIM IMPLEMENTATION AND EVALUATION

4.1 Preliminary BIM implementation Framework

The literature review included BIM implementation documents, such as BIM standards, execution plans, protocols, BIM implementation frameworks and assessment tools. By classifying and renaming of terms relevant to BIM implementation, this research firstly developed a preliminary BIM implementation Framework (BiF). In a multi-structured level (Lee and Burnett, 2008), it contains 69 criteria which can be divided into five dimensions, which are ***project management, data management, application management, organizational management and stakeholder involvement***, as shown in the Figure 4-1 overleaf. With regard to the purpose of this research, this BiF could facilitate industry awareness for key criteria of BIM implementation and help BIM management issues (Chien et al., 2014). Considering the wide coverage of the literature review, the developed BiF intended to be generic and to be used with a wide range.

These five dimensions as aforementioned has been relatively independent, which allow user to evaluate BIM's capability in each individual dimension e.g. project management, organizational management, data management, application management and stakeholder's involvement. The summation of each dimension's maturity could reflect the overall BIM capability on the entire project. The preliminary developed BIM implementation framework contains five dimensions and 69 factors, which have been further refined via case studies and questionnaire as explained in the following sections.

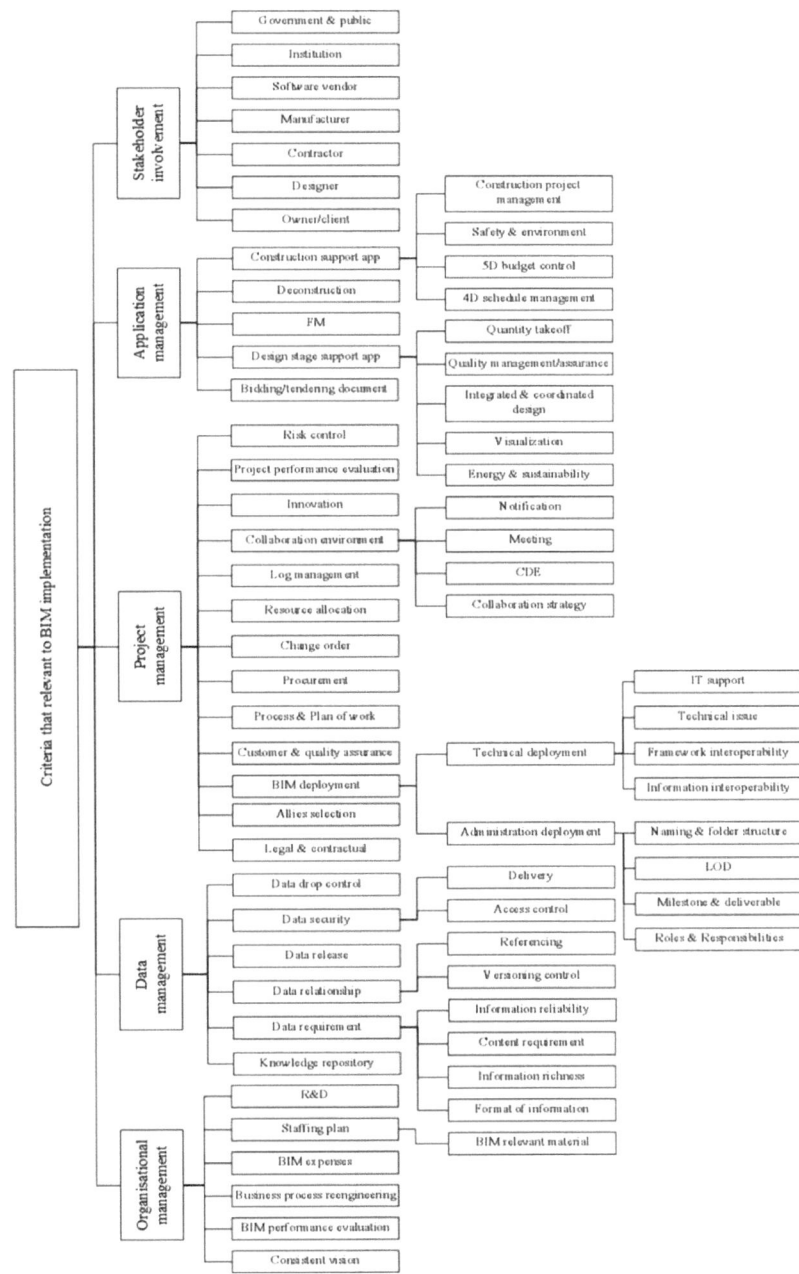

Figure 4-1 The Preliminary BIM implementation framework

4.2 Framework enhancement through practical case study

4.2.1 BIM pilot project

The internal pilot project was conducted within the BIM research group, where researchers played different roles according to their expertise, such as BIM manager, architect, structural engineer, service engineer and so on. The purpose of this pilot case study was to accumulate experience and solve potential problems during the BIM deployment process, hence to prepare for the industry project that has been selected for BiF validation stage.

The author played a role of BIM manager to manage the collaboration between stakeholders, data management, and architectural model design. Other researchers took other roles, e.g. structural engineer and service engineer. Criteria within project, data and application management dimensions have been adopted, such as: naming & folder structure, 3D modelling, multi-disciplinary collaboration, data access control, 4D scheduling modelling and clash detection. In addition, problems during the adoption process have also been summarized and used to further improve the developed BIM implementation framework.

4.2.1.1 Collaboration environment set up

Firstly, the collaboration environment among disciplines has been created. All stakeholders should be able to access the central database constantly, which also requires a well-controlled data management policy to guarantee the safety of data. BIM also requires integrated and collaborative design platform among all stakeholders. The selection of BIM authoring tools therefore should consider issues like compatibility, visualization, integrated design among disciplines etc.

Revit BIM authoring tools provided by Autodesk have been used for this case study. Two options for data storage and exchange have been tested. Shared Network Drive (SND) was used for internal data exchange. All workstations in the local area network were connected to an Ethernet switch with total transmission speed of up to 1 GB/s. With the consideration of cross organizations activities, Dropbox was selected as an extra central repository where all data was located and shared among all external stakeholders. A proper folder structure and naming convention has been applied as well (AEC (UK) Initiative, 2012).

Revit: Worksets and Link Model

Other than essential modelling capability, Revit also offers Worksets and Link Model functions to facilitate data management and collaboration among users.

When the Revit model file was first created, a Central Model was saved in the Dropbox shared folder. Each designer who would contribute to this model would create a Local Model on their own workstation. Data belong to the same person controlled by one Worksets to avoid unauthorized access, while others can edit only with the owner's permission. Revit allow designers to upload their changes made in their own Local Model to the Central Model, the Central Model will then notify all other Local Model owner that an update is available for them to synchronize with the latest information.

There might be only one Central Model for each discipline, Revit also allows different model to link each other as references. This brings many advantages, such as automatic model coordination, change notifications and model segregation. The advantages are: first, other disciplines e.g. structure engineer could link architectural model in an earlier stage (e.g. after columns and walls etc. have been placed in position) for them to start structural design earlier with concurrent effort (Fischer, 2006); secondly, if there is any changes from architect side, structure engineer can be notified immediately after the Central Model has been updated.

Navisworks: clash detection, 4D BIM

After all design work has completed, all disciplines' model would be exported into NWC format which is compatible by Navisworks. Navisworks is a coordination software by Autodesk. It overlaps/merges different 3D models to detect any conflicts or clashes. For example, different pipes can accidently cross each other, which is usually hard to identify via traditional approaches.

After all design models have been reviewed for potential clashes, another functionality of Navisworks has been adopted: 4D BIM, the scheduling management by integration of the construction schedule with the design model. The capability of simulation could present the sequence of how the building can be constructed from scratch; this aligned with the construction plan provided from contractor. Moreover, costing information can also be presented for user to monitor the budget information at any point of the construction activities within the virtual environment.

4.2.1.2 Practical adoption issues and solutions

Other than the basic key implementation areas that have been applied in this case study, there are also a number of technical issues which have been identified. In Revit, designer will have to select the expected 'family' to produce the actual building component. However, projects will be different from each other, while there is only limited number of default family for designer to select. The designer would either create a new family to meet the client's need, or use one of the default types to replace their expected one. But both of them could cause problem in future: if the family was not created properly, it could affect quantity take-off (an automated process embedded within Revit) or even cannot be recognized by manufacturer. If a replaced family has been used, more confusion could happen during later construction stage. Moreover, family such as elevator could only be created by highly experienced and skilled designer or even manufacturer.

When the Worksets have been borrowed by another designer with permission of its original owner, the designer have to release the ownership otherwise other users will not be able to use it.

4.2.2 Real-life industry case study

4.2.2.1 Project introduction

The second case study was selected based upon several reasons: (1) a typical building project so it could represent other building projects; (2) the project requires multidisciplinary interaction during the detail design stage as a minimum requirement, and (3) no BIM implemented. All these factors provide a 'test-bed' for the proposed BiF. A deep understanding of the current workflow is required which is essential for the author to develop a tailored BIM plan according to the needs of the local group (Hartmann et al., 2012).

Ethnography and interview methods have been used for this purpose (Oates, 2006). Any outsider involved in an alien circumstance/environment
to study its culture and behavior is termed as an ethnographer. The present author joined the pilot industry organization and acted as a visitor and learner, invisible to the local employees. By conducting such an approach, it allowed the author to find out first-hand reliable information of current industry BIM practices and its workflows; and how people there work, communicate and collaborate. Moreover, interviews with selected heads of each discipline were also carried out to understand comprehensively about their specific targets, processes, management and workflows.

In order to best demonstrate the efficiency of the proposed BIM implementation Framework (BiF), an organization with no BIM background was selected to conduct a real-life project based case study.

The organization was interested in BIM and was willing to fully invest on BIM, including hardware, software and relevant organizational changes.

The selected organization was SAIXA (anonymous), established in 2008. SAIXA has an integrated design department including Architectural team, structural team, construction drawing production team and operation management team which was in charge all the administration work.

This project selected was a two storey steel structural office building in China. The lifespan was designed for 50 years, Seismic Intensity Protection for 7.5 degrees on the Richter scale (UPSeis), and a construction area of 1500m^2. The client was one of the New Developing District, a national level district of China which requires high level of sustainability in design.

Since that was SAIXA's first BIM based project, they started the BIM process with the training process at the same time. The case study focused on the detail design stage, where the collaboration and data exchange mainly occurred. For the comparison purpose and contingency plan, SAIXA also appointed a second team to apply the traditional approach on the same project, in parallel. This was also used to help them to evaluate the actual benefits of BIM in practice. The case study was carried out from 9th July to 25th August 2011.

4.2.2.2 Case study design

In order to have full support from the organization, firstly an individual meeting with the owner of SAIXA was arranged. The present researcher explained the research purpose and plan, especially what SAIXA could benefit from this study. The owner agreed that the researcher could have his full support in terms of human resources and financial support.

A group meeting was then conducted, all department leaders were asked to attend. The presentation covered: a brief introduction of what is BIM and how BIM could assist with each discipline's business activities. Individual meetings with each department head were arranged, and that allowed the researcher to observe the current practice at SAIXA using the developed BIM implementation Framework (BiF) as a reference to propose the BIM strategy. The BIM strategic planning was then applied in one of their real projects led by the researcher. Finally, the feedback was collected from the users to further improve the developed BiF.

Each team member were also interviewed individually. The team includes: an executive director, an assistant of the executive director, head of concept design team, head of architecture and construction drawings and mainly responsible for review and approve design work, head of structural design team, a senior plumbing engineer, a senior electrical engineer, a senior HVAC engineer and an administration officer.

4.2.2.3 Observation for existing practices

Existing Design platform

SAIXA is currently using Tangent software (A 2D model development software build on the platform of AutoCAD: http://www.tangent.com.cn/) for design work. With comprehensive libraries that aligning with industry standards, Tangent maximum reduce designers' work load. However, there was no clash detection function available, all clashes can only be detected during construction drawing production or construction stages.

Collaboration environment

All project information were stored in personal workstations. There was no backup policy for the safety of information. The communication and collaboration was very traditional. The exchange of information across disciplines within the organization, was based on an instant messenger; the exchange of information across organizations with other stakeholders, was also based on instant messenger or email. Since there is no 'published' zone as explained in Common Data Environment (BSI, 2013), all information transferal could only be done by the head of the design department to ensure a correct information to be delivered to the client. An incomplete drawing cannot clearly demonstrate design intention and to support client's next step of work, e.g. contractor selection. This could result in workload accumulation for senior engineer, and delay the project delivery time and decrease the quality. All available working approaches are uncontrolled, as there is no record of the exchange of information. This could result in dispute when a superseded version of drawings have been produced.

Existing work process

Concept design: the concept design team was responsible for developing project proposal for client's approval. The proposal included project functions, construction area, similar business cases analysis and proposed 2D plan views and 3D model for demonstration etc. However, the 3D model is normally based on SketchUp or 3Ds Max, which cannot be continuously used in later design stage. The completed project proposal was then sent to City Planning bureau in 2D format for approval.

Detailed design: after the concept design, architects continued to develop 2D drawing in all views (e.g. plan view, side and elevation view and cross-section view) in a more detailed level. After the completion of the main components of the building, drawings were provided to other disciplines as a starting point for their own work. Different scales of drawings were continuously produced by architect (e.g. 1:100 for overall view; 1:50 for kitchen, staircases; 1:20 for slab detail; 1:10 for roof drainage eaves etc.), which could not be generated automatically. During this process, the architect had to initiate several internal meetings with other disciplines to discuss potential issues and other's discipline requirements. Modifications are expected at all times, along with a large amount of iterative work.

In the meantime, the structural engineer started loads design including point load, line load, area load, external loads etc. Electrical engineer asked for information regarding power and heating requirement from the client and produce drawings based on architect's drawing. Plumbing engineers asked for relevant information of sewerage output and domestic water input.

All teams were fragmented, work process was deprecated: communication was delayed due to relevant stakeholders not being notified immediately or automatically regarding recent changes, thus low efficiency was unavoidable. All disciplines provided their requirements to the architect to update his design, e.g. to rearrange space functions regarding machine room, power supply room and to reserve holes on the wall for electricity wires and piping system etc. Because not all disciplines were involved since the concept design stage, all these led to rework of architectural drawings, recalculation of loadings, all of which would need the client's re-approval. This also caused information losses, inaccuracy and incompleteness. By the end of developed design stage, the completed architecture drawings were then provided to structures team; additionally, cross section view of building components (e.g. column, wall, slab and staircase etc.) were produced to demonstrate their internal structure: reinforcement and steel bar arrangements.

Construction: after the construction drawing have been completed, the drawings would be delivered to the client and the client would appoint a contractor. Since the 2D based MEP drawing is complex, in order to speed up the construction process and avoid confusion in reading drawings, a question and answer session was organized between designer and the contractor before the actual construction started. This also provided assistance for construction strategy development. During the construction stage, regular site visits were arranged, designer has to make sure the actual construction follows plans. Coordination meetings would be proposed if any changes were required. Such changes could be required by the client, due to lack of collaboration between design teams in earlier design stage or limited by the contractor's capability (for example, lack of relevant skill, high cost or high risk). The designer would have to re-design and re-submit for government approval. This could lead to serious construction delay.

4.2.2.4 BIM strategy planning and implementation

After the observation work has been completed, based on the current practices and readiness, the researcher has developed a BIM deployment strategy - the proposed BIM implementation Framework (BiF) was introduced to the team including how BIM should be implemented in an organization; what are those implementation criteria; how data should be managed under BIM environment; what are the main applications of BIM, and which should be prioritized in the project.

Setting Up the BIM Environment

The working environment is the foundation to allow BIM to be deployed with expected performance. This will mainly include hardware, software and data storage requirements.

Equipment & Facility

The concept of BIM is realized through the use of BIM authorizing tools for modelling, coordination, collaboration and project management etc. Such activities require a powerful workstation especially for modelling and rendering functions. Three workstations with Intel i7 CPU with 3.4GHz speed and 8GB for memory were purchased especially for this case study. Revit 2012 series and Navisworks 2012 have been used for

BIM modelling and coordination work.

1. Revit Architecture for concept design, architecture design and energy analysis export tools;
2. Revit Structure for structure design;
3. Revit MEP for HVAC, electricity and plumbing design;
4. Navisworks for 4D simulation and clash detection.

Data management: modelling and storage

In this case study, LAN based shared drive was established to store the Central Model, as all data was mainly exchanged internally during the design stage. All project documents were saved in an appropriate structure and with a specific naming convention according to BIM standards (AEC (UK) Initiative, 2012), with an appropriate storage system (WIP, shared, publish and archive) for convenient document management work. User access was controlled using appropriate ownership rules.

There were two architects working on the architecture model. Worksets function worked well in this case to prevent accidental component deletion. *Send Request* function was used to allow change in element ownership. The original designer would grant permission, and will be informed of changes that have been made by others, for review and approval. This effectively reduced potential disputes caused by model ownership. *Link Model* enables designers from different disciplines to link each other's models as reference to support coordination.

The communication within BIM team is still based on its commonly used methods, such as LAN communication tools, instant messengers and face to face meeting. However, due to model liking, worksets and notification mechanisms provided by Revit, the demand on communication has been largely reduced. Design solution could be consistently and quickly exchanged between members, and that improved their work efficiency. The researcher conducted several training sessions to help designers to use Revit software.

Organization Management

The existing organization structure would have to make changes so as to meet BIM requirements. In order to avoid unexpected issues, a new department was established first (Tsai et al., 2010) - its members are the head of each design discipline and the operation department. The researcher took the role of BIM manager to manage BIM based project and reported to the management team about progress and achievements.

BIM based Design Workflow

After the completion of the concept design stage, both teams started at detailed design: the BIM team worked on Revit platform while the other team followed their normal method. In order to maintain a consistent design between two design teams, the same design was used, such as loading system, piping system and electrical system etc. Therefore, the main differences between these two teams were: modelling environment, working process, collaboration method, data management and applications etc.

Set up Project Milestones

The project manager created a schedule (Table 4-1 overleaf) which included: deadlines, main activities and associated model requirements. The entire project progress was monitored by Project Manager and Client. Reasons for delay have been recorded along with the adopted delay solutions.

Data management

The BIM Manager or Project Manager was responsible to set up a standard for data requirements. All BIM models were created in Revit, their formats have to be compatible with other users. Within the model, other than traditional graphical information, non-graphical information was also required, such as price, supplier information, other comments etc. The model was developed comprehensively to meet the requirements for internal and external reviews.

Project Activity	Dead Line	Main Activity	Model Required
Project Mobilization Meeting	26th July 2012	Appoint Project team member and leader, assign tasks and delivery information	Concept Design Model
Construction Site Drawing	27th July 2012	Construction Site Planning	NA (already completed)
Architectural Model	2nd Aug 2012	Produce 3D Model based on	Revit Architectural Model
Material Selection	3rd Aug 2012	Brick and Concrete Supplier Procurement Path	NA
Structural Model	8th Aug 2012	Produce 3D Model	Revit Structural Model
MEP Model	15 Aug 2012	Produce 3D Model (Electrical, Plumbing)	Revit MEP Model (Electrical, Plumbing Design)
Clash Detection	17th Aug 2012	Clash Report and Model Update	Navisworks Management Model & Clash Report
4D Model	20th Aug 2012	Construction Scheduling	4D schedule simulation
Submission for Approval	25th Aug 2012	Submit Design Work to Government for Approval	All BIM model will be Submitted

Table 4-1 Project Milestones of real project in SAIXA

When a new model version was created due to a client change or conflict with other disciplines, the old version was then archived, and a note was attached explaining the reason for change. All changes were agreed and approved by head of department and available to all stakeholders immediately.

The access history and actions were recorded and maintained consistently, therefore Revit can automatically export this information if there is a need to review operation history.

BIM application

Energy & sustainability: from project start, Revit could be used to develop 3D model for Energy Estimation. Environment method such as BREEAM could be used to improve the sustainability of the design. Solar study could also be conducted to help the designer to evaluate the impact of the natural light and shadows on the model in different seasons. All these should have done in an early design stage for the client's reference to come up with a more optimized solution. However, the concept design stage of this project has already completed, and there was no requirement on the energy efficiency or sustainability aspect.

Integrated design: After the initial architectural model was completed, other disciplines such as structural engineer, MEP engineer used 'linking' function in Revit to link with architectural model and copy essential element to demonstrate the design intention of the project, and then to started their own design works.

The developed integrated design environment allowed BIM based collaboration. If anything change in the architectural model, structural and MEP engineer would be automatically be notified by Revit Link Model system; if the structural engineer would like to make any changes on the position of the architectural element, he could borrow the ownership from the architect, to make changes and that will be reviewed and approved by the architect. Architect would continue the design, e.g. to allocate functions to each room, place windows and doors etc. Structural engineer would export completed structural model to Robot Structural Analysis for further calculation. Plumbing engineer will design Domestic Water Supply, Sanitary Waste and Vent and Storm Drainage system. Electrical engineer would design Lighting System, Switch System, Power System and Panel board System. All components will be listed in a new data sheet for quantity take-off purpose.

Clash detection: all design work completed by Revit was exported into NWC format and imported into Navisworks, where clash detection can be conducted among all discipline models. In this project, seven clashes have been identified relevant to the light fitting and the wall.

4D schedule management: after all clashes have been solved, the models were used for further application, a planned construction schedule was designed and applied in Navisworks to demonstrate the construction procedures.

Budget control: the costing information were embedded within the BIM model, and included within the quantity sheet produced automatically by Revit. Relevant information was provided by contractor. Navisworks was used to perform a more detailed costing information: material cost, labor cost, equipment cost, subcontractor cost and total cost. All these information were presented separately in Navisworks schedule simulation and accumulated along with the project progress.

4.2.2.5 Result and feedback

After the case study has been completed, each member of the BIM team submitted a short paragraph on their experience in using BIM and advantages from their own discipline. Main advantages of BIM compared to traditional approach has been shown in Table 4-2.

There are also a number of BIM implementation barriers concluded during the project case study, e.g. (1) the use of BIM concept within design organization requires significant initial investment. While there are no feasible existing decision making tools to guide the implementation process; (2) There is need for a consensus among all project participants regarding the focus scheme, and therefore to have BIM strategically implemented; (3) The quality/level of using BIM could not be measured. Adoption issues need to be identified. The adoption of BIM should align with the organization's vision to maximize BIM's advantages.

Practical performance indicators	Description	Group 1: Non-BIM approach	Group 2: BIM based approach	Advantages of BIM
Collaboration	Data communication among various disciplines	-2D based -Fragmented and low efficiency	-3D + non-graphical information-Collaboration with others but limited due to the first time using BIM	Improve the information exchange efficiency within and cross disciplinary
Project schedule management	If the project progress is following the project milestone as planned	Information delivered before the actual deadline but with large effort	Information delivered before the actual deadline with less modelling effort	Reduce work load, modelling time and waiting time, avoid re-work, easily to generate 2D views
Paper consumption	Drawings will be printed for all meetings or discussions	Missing or incorrect information require re-print	Information is correct and complete	All information can be reused, reduce incorrect or incomplete

				information

Practical performance indicators	Description	Group 1: Non-BIM approach	Group 2: BIM based approach	Advantages of BIM
RFIs management	Information request for any incorrect or confused drawing	2D based drawing could result in understanding issues in construction stage	3D model assist engineer understand drawing	Demonstrate design intention through both 2D views and 3D model
Clash Detected	Clashes that been detected during construction drawing production stage	Clashes cannot be identified but remain until construction stage	Clash has been detected in earlier design stage	Avoid clash in later stage, especially in actual construction stage
Sun Lighting/visualization	Analyze of sunlight	Unknown the effect of sunlight	Proposed design solution can be virtually visualized and sun path simulated	Improve client and designer's visual impression for a better design solution

Table 4-2 Performance comparison between BIM group and traditional group in case study in SAIXA

4.2.3 Framework revise based on expert consensus

As discussed before, three round of Delphi expert panel had been conducted to collect professional BIM users' opinion. The panel selection is the decisive factor for success (Jillson, 1975, Hasson and Keeney, 2011). Therefore a strict policy on the selection of organizations and participants was proposed and followed in this research in order to collect the most comprehensive, reliable and convincing data (Chien et al., 2014, Meesapawong et al., 2014).

Six prestigious design and BIM expert consultant organizations were invited, according to the following criteria:

1. Each organization should have a minimum 3 years' experience of adopting BIM; in fact, the selected organizations are the first users of BIM and have contributed to the development of China BIM standards;
2. BIM based project experience in all types of project e.g. housing, health care, corporate, sport, education, etc.
3. Their roles and responsibilities mainly cover the design stage but still maintain a close interaction with other stages (e.g. construction & operation stage).
4. Geographical difference to reduce perception and phenomenon affected by localisation; all organizations as well as their branches are located in different regions of China, while their businesses are all over China and mostly globally, therefore to show an average phenomenon of China;
5. Multiple types of organizations were selected to collect perception from multiple aspects; such as global design and consultant organization, local design and consultant, government owned design and research institution etc.
6. The organization should have diverse nature of disciplines to collect various experience and perception from individual

The selected participants play different roles in their organizations:

1. Director/CEO and technical director: has a unique view, control of BIM development in the company and a long term vision of BIM in China's industry;

2. BIM manager, BIM expert and BIM senior engineer of the design team who may have consistent knowledge, expertise and experience in BIM in practice;
3. BIM engineer or experienced consultant who could have a better knowledge and experience in BIM implementations.

By the end of Delphi study, initial BiF had been revised (Figure 4-2): four removed factors has been removed:

1. *Allies selection*: it aimed to have partners with the most adequate BIM capabilities on board for project completion with an expected target. Participants argued this was duplicated with the stakeholder involvement dimension. Additionally, the client would usually be responsible for this criteria;
2. *Procurement*: its purpose was to provide a direct answer to the general question of how the project can be done (Porwal and Hewage, 2013). 'Procurement' normally will start a tendering process. When the client is looking for a designer, the contractor or other supply chain partners, it would give client's requirement to the candidates. Therefore its focus is on the process of the work that is being done, as well as how the work should be delivered. In other words, procurement duplicates with terms e.g. contract, process, quality and data management aspects, hence it was removed from the framework.
3. *Framework interoperability*: this term turns out to be particularly important when considering issues related to sensor network (Shen et al., 2010). However this is mainly considered during the operational phase. Hence it received a lower credit from the panel.
4. *Notification*: aimed to inform stakeholders of information update or project schedule approaching. However, such

function has already embedded within existing design platform as an attached function.

5. *Delivery*: the delivery of data in BIM circumstance is mainly through BIM server or cloud based, which mainly rely on high security of protocol. Hence participants believe this criterion overlaps the *access control, collaboration strategy* and *legal and contractual issues* etc.

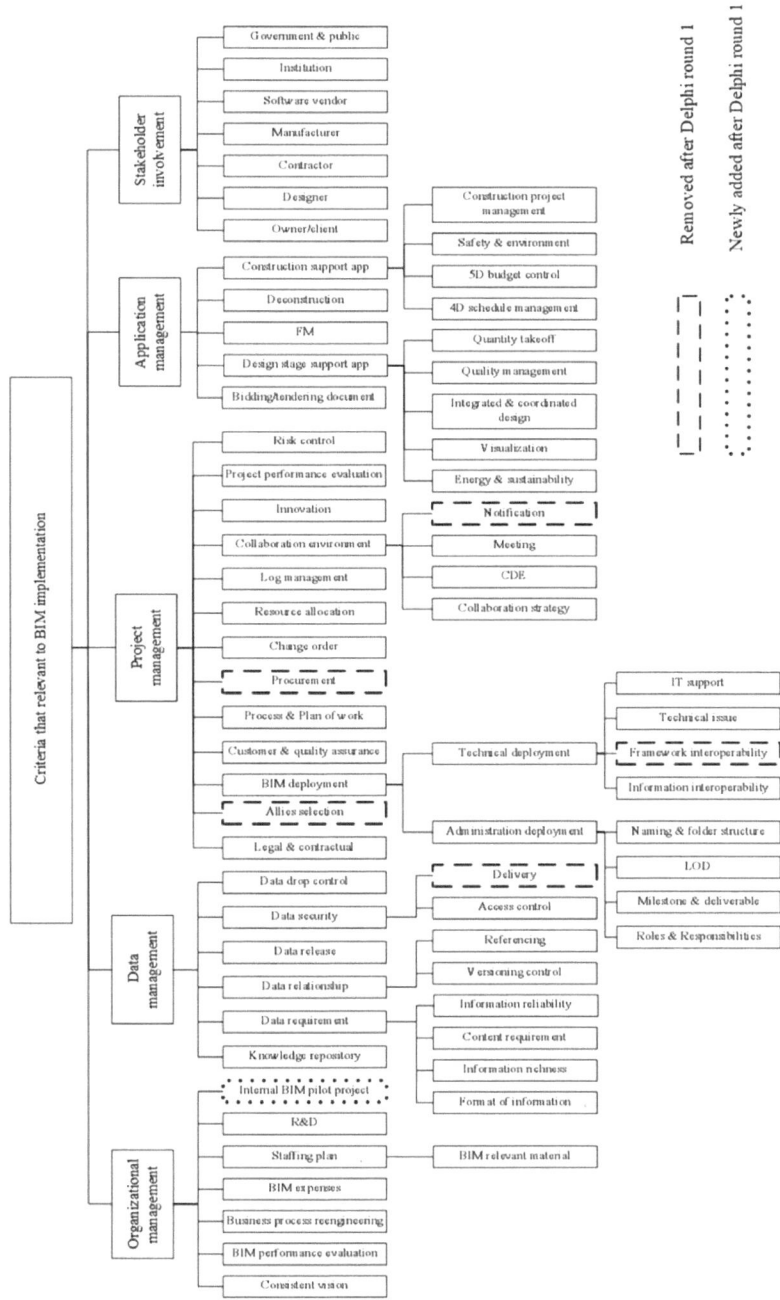

Figure 4-2 The proposed BIM implementation framework by Delphi study

One factor was added: *Internal BIM pilot project*: before the actual deployment of BIM, it is best for BIM users to carry out several pilot projects with other partners (Howard and Björk, 2008, Porwal and Hewage, 2013, Tsai et al., 2014c). This could be conducted internally or small scale project, to simulate real situation and identify any potential risks and solutions (Zahrizan et al., 2013). This is especially necessary to get familiar with new form of business structure and process. This was agreed by all participants to improve the final project outcome, and most importantly, to increase their confidence and willingness during the transformation process.

Therefore, a fully functional BIM implementation framework has been completed and is ready for wider industry trial.

4.3 Summary

In this chapter, a consensus based group decision making method: Delphi method was employed to further refine the result from previous research steps in a specific context. The final BiF contains five dimensions and sixty five factors that most relevant to BIM implementation in China.

However, Delphi method only delivered a set of refined factors without ranking the proposed factors and as such do not possess any practical implementation value (Meesapawong et al., 2014). To complement this, the AHP method was adopted in the next step. AHP as a group based decision making method which will intake the criteria concluded from Delphi method, and create a strategic decision making tool for organization to deploy BIM thoroughly..

5 AHP BASED STRATEGIC DECISION MAKING FOR ORGANIZATIONAL BIM IMPLEMENTATION

5.1 Analytical hierarchy model development

The purpose of this hierarchy model is develop a tool that can be used by the management team to make decision for BIM evaluation and strategic planning for future development. Required by ASZ, the decision making criteria demands a managerial perspective.

The final approved hierarchy framework as shown in Figure 5-1: four dimensions have remained as the second level of the hierarchy framework (H2). Level three (H3) comprises 21 criteria while level four with 14 sub-criteria. Moreover, the management team agreed that, in order to deploy BIM in a strategic way, at least one focus scheme was needed to guide the implementation direction. In ASZ's case, three focus schemes have been proposed and located at the last level (H5) of the final hierarchy framework. Their description details are shown below:

1. *Sustainability* is one of the most important topics in the global AEC industry (Wong and Kuan, 2014), especially in China (MOHURD, 2013). This could be achieved by improving environmental considerations (e.g. LEED, BREEAM etc.) to reduce energy consumption from the early design stage and focusing on the social welfare, to consider the environment, low carbon emission and eliminate unnecessary construction reworks during construction stage.

2. *Commercial value* aims to deliver the most profit and commercial value out of the services to the client and the products. It also enables convenient decision making for the client, especially in the preliminary design phase. Data-based project management also simplifies the developer's workflow. Moreover, marketing can now be integrated into design in a very persuasive way.

3. *Customer satisfaction:* The 'customer is the most important part of the production line' (Neave, 1987). As one of the tangible benefits of BIM (Arayici et al., 2009, Khosrowshahi and Arayici, 2012, Ali et al., 2013), a better customer appraisal scheme could maintain a good rapport with clients for repeated business: (a)for the investor: shorten the schedule and budget without any compromise; focusing on the quality of the product, such as energy simulation to achieve operating expenditure economize; (b) for the user: emphasize on the comfort experience of building users by helping perform wind and floor vibration test, for example.

The proposed three options will be the future focus scheme of ASZ, hence the impact of each of them towards each criterion will based on their current knowledge, understanding and attitude.

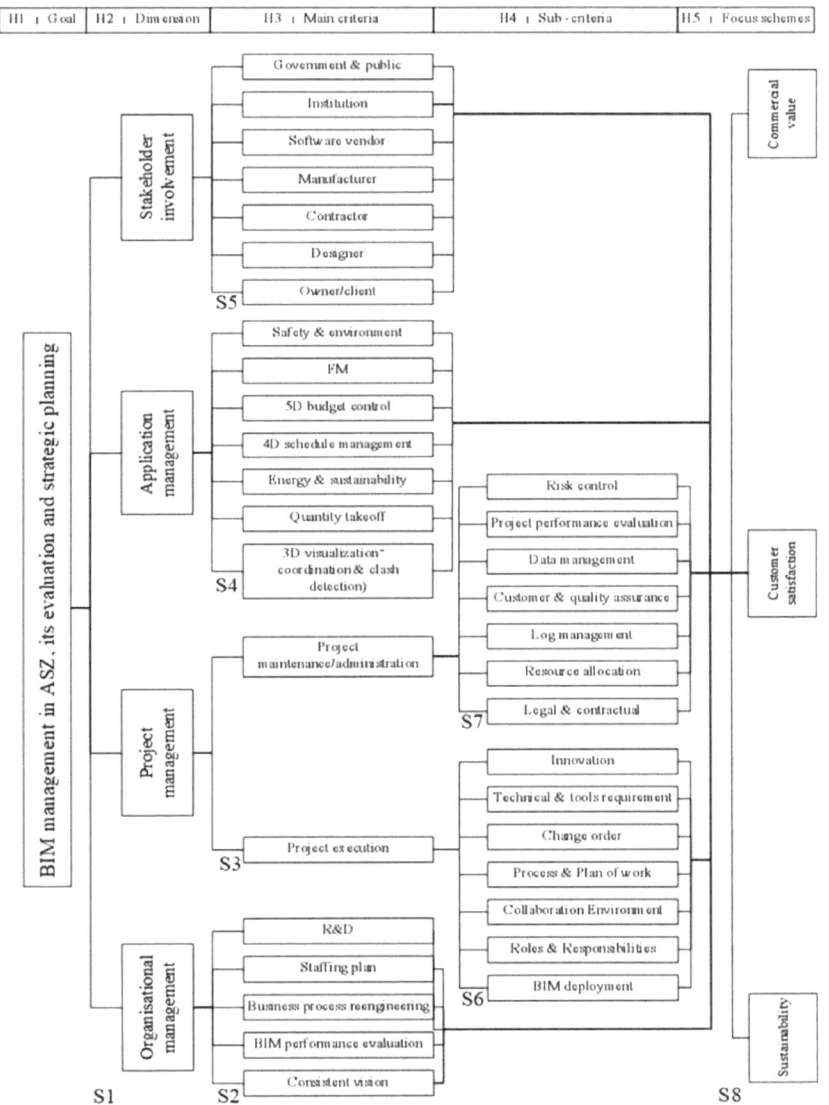

Figure 5-1 Hierarchy model for BIM implementation in ASZ

Hierarchy Level		Global (%)
1	Goal: Future objective orientation of modern AEC organization, a case study of ASZ	-
2	Project management	26.7
3	Project execution	9.9
4	BIM deployment	2.4
4	Roles & Responsibilities	0.9
4	Collaboration Environment	1.4
4	Process & Plan of work	1.6
4	Change order	1.3
4	Technical & tools requirement	1.1
4	Innovation	1.2
3	Project maintenance/administration	16.8
4	Legal & contractual	2.3
4	Resource allocation	2.2
4	Log management	1.2
4	Customer & quality assurance	3.1
4	Data management	1.6
4	Project performance evaluation	3.1
4	Risk control	3.3
2	BIM application	8.4
3	3D visualization, coordination & clash detection	0.7
3	Quantity take-off	1.5
3	Energy & sustainability	0.5
3	4D schedule management	2
3	5D budget control	2.5
3	FM	0.7
3	Safety & environment	0.5
2	Organizational dimension	50.6
3	Consistent vision	3.4

Hierarchy Level		Global (%)
3	BIM performance evaluation	25.3
3	Business process reengineering	12.3
3	Staffing skill, plan & training	6.3
3	Internal R&D	3.4
2	Stakeholder involvement	14.2
3	Owner/client	5.1
3	Designer	2
3	Contractor	1.2
3	Manufacturer	1
3	Software vendor	0.7
3	Institution	0.7
3	Government & public	3.4

Table 5-1 Result of AHP study: global and local priority and CR value

The AHP Table 5-1 shows the result AHP result represents the weight of each factor.

5.2 The weighting system development

As shown in Table 5-1, general speaking, for the weighting system development, a higher global priority implies the specific criteria has a higher weight during the implementation stage, which will be considered to be implemented first when there is a resource constraint (Jung and Gibson, 1999).

The result shows the *organizational management dimension* received the highest priorities (50.6%) which is aligned with existing research findings (Penn State University, 2010). *BIM performance evaluation* (25.3%) received a high attention in their routine business activities to evaluate, control, predict and therefore to ameliorate the critical factors that affect project performance (Luu et al., 2008). Under BIM philosophy, conventional *business process* require *re-engineering*

(BPR 12.3%) to integrate all business processes and projects within the organization. Therefore, all individuals can get involved (Muthu et al., 1999) to improve strategy advantages. While managerial benefits still need to be quantified and clarified (Jung and Joo, 2011). Employee's awareness and confidence in BIM can be improved through *learning and training scheme* (6.3%) (Pickford, 2013). Training will vary with the roles and responsibilities, and people in management position require a different and specific training (Singh et al., 2011). This will also enhance their competitiveness and enthusiasm in using BIM (Eadie et al., 2013).

Project management dimension received 26.7% of focus for BIM development, where *project maintenance/administration* (16.8%) received much higher attention than *project execution* (9.9%). With the consideration of the priority of *organizational* dimension, the numeric data obtained implies ASZ were more focused on the management development of BIM. *Risk control* (3.3%) can be further divided into two aspects: to reduce risk caused by adoption of new technologies e.g. software malfunction and information inaccuracy (Porwal and Hewage, 2013); and meanwhile to avoid project failure (such as ownership, legal risk (Mahalingam et al., 2010)) by relying on the new technology (Shafiq et al., 2013). *Customer & quality assurance* (3.1%) requires to reach the customer's satisfaction (e.g. Quality Insurance, schedule and budget control of the project) and other project performance evaluation perspective (e.g. environment, health and safety policy). Key performance indicators (KPI) are normally used to measure the *Project performance* (3.1%) of specific task of BIM throughout the project (Barlish and Sullivan, 2012).

Stakeholder involvement dimension received a slightly lower percentage: 14.2%. This reveals the value and effect of stakeholder during the building lifecycle still haven't reflected on the actual project. In the current AEC industry in China, the use of BIM is still the *client's decision* (5.1%), since the client is responsible for the selection of designer and contractor and has the pull to enhance the project performance as well as to pay the additional. *Designer* (2%) still dominant the BIM implementation process by starting the BIM model, and continuously inputting information into the model and passing the model to downstream users e.g. contractor. in order to improve industry-wide BIM competencies (Succar et al., 2013), all other partners should have their own roles therefore to form an entire BIM loop from a lifecycle view of building (Mahalingam et al., 2010). Especially the involvement of *government* (3.4%) could propose associated policies, standards and guidelines for BIM, as well as mandatory requirement for all project to meet a certain level of BIM usage rate and sustainability rate. *Manufacture* (1%) provides BIM object library to designer to ensure the information within the BIM model is compatible and meets the requirement of the protocol (NBS, 2013a).

Application management receives the lowest attention from industry (8.4%). In one side, BIM technology development has been well developed (Jung and Gibson, 1999) therefore requires less resource input; in another side, specific application is more about technique adoption which has less influence in term of management e.g. modelling, simulation and analysis. The data shows ASZ currently has a focus on *5D budget control* (2.5%), *4D schedule management* (2%) and *quantity take-off* (1.5%). *4D schedule management* has a CR of 9.6%, which indicates participants believe it has a less obvious influence on various focus scheme.

6 BIM EVALUATION FRAMEWORK (BEF) DEVELOPMENT & VALIDATION

6.1 BIM evaluation Framework (BeF) development

The AHP method delivered a weightage system, which identifies the priorities of each criterion in the analytical hierarchy model, as shown in Figure 5-1. Therefore, an framework for BIM implementation evaluation during the project design stage is ready to be developed (Sebastian and van Berlo, 2010). Based on the aforementioned research steps, eleven criteria have been preliminarily concluded for BIM implementation evaluation:

1. Coverage: the assessment scope of the framework should focus on all the dimensions of BIM implementation instead of single aspect e.g. data management aspect;
2. Reliability: the assessment criteria adopted by the framework should come from a reliable source e.g. official BIM standards;
3. Applicability: the applicability of the selected assessment criteria should be proved by empirical evidence;
4. Qualification: the selection of BIM users for criteria verification should follow certain requirements e.g. disciplines, experience in BIM etc.
5. Consensus: the group consensus for each assessment criteria should be achieved;
6. Weightage: the weight of each factor should be calculated scientifically and should be obtained based on all project stakeholders' opinions;

7. Prioritisation: the criteria should be prioritised and the sum of all weights should be 1, instead of using an average value or a simple ranking order to represent their priority;
8. Focus scheme: the development of the framework should consider the organization or project's focus scheme;
9. Ranking order: the ranking order of each focus scheme should be obtained;
10. Impact: the impact of each criterion on various focus schemes should be clear (e.g. either a positive or negative influence) and therefore improve the organization's attitude on the selected focus scheme;
11. Benchmarking: the framework should have the potential to be developed as a benchmarking method for comparison among organizations.

In order to develop an evaluation framework, the following steps have been followed in this research:

1. Based on the management team's requirement, the original BiF had been modified and all criteria contained in the hierarchy model used in AHP step was more about the managerial perspective. However, the use of BIM in project is multi-dimensional, which includes all aspects including detail level such as technical implementation. Hence a connection between managerial criteria and other criteria as included in the final BiF. The weightage to every single criterion was allocated. However, there were some changes on the criteria based on the demand from ASZ:
2. To quantify each criteria (Najafabadi, 2013) into several scales/levels, to describe the progressive improvement of BIM usage from non-BIM to collaborative BIM of that particular

criteria. The weight of each criterion will be equally assigned to each level it can be divided.

The proposed BeF has the following applications:

1. To be applied by project manager or BIM strategist at the **beginning of the project** to assist with decision making and BIM implementation strategy based on the prioritised criteria:
    - To decide priority of potential BIM applications; and consider their interrelationships and impact to client's business case in long term;
    - To define responsible party and deliverables etc.;
    - To assess stakeholders' BIM competencies and available resources;
    - Predict risk and revenue etc.
2. To rapidly review the ongoing project's condition and come up with the best option to optimise the follow-up process and activities **during the project lifecycle**.
3. The closing process of a project will guarantee the formalised acceptance of a project or a certain phase and lead to an orderly end (Kwak and Ibbs, 2002). Kwak and Ibbs also argue it would be useful to develop a systematic lessons learned documentation, therefore the developed BeF could perform a **closing inspection** to discover weak and strong points from the completed project for future improvement.
4. To evaluate the best BIM based project that a company has completed, therefore to obtain its BIM capability.
5. Designer prequalification decision making process (Russell and Skibniewski, 1987, Porwal and Hewage, 2013) can be performed: the client will assess the candidates' (potential design team) BIM capability by using the proposed BeF,

therefore to eliminate less favourable candidates: those who has received a relatively low evaluation result (Sonmez et al., 2002, Sebastian and van Berlo, 2010).

The developed assessment framework has been partially shown in Figure 6-1. The assessment framework was developed by programming in Microsoft Excel.

Column A-C displays all assessment criteria; their associated definitions are displayed in column D-E, to ensure there is a clear understanding regarding each specific factor (Jung and Gibson, 1999).

The maturity model requires to have a set of incremental levels to represent the involvement of BIM (Succar, 2009b). According to some existing literature (Rezgui et al., 2013, Succar, 2009a, BIMIWG, 2011), the application of BIM can be categorized into three levels:

1. Level 0: traditional CAD based approach;
2. Level 1: mixed 2D CAD and BIM approach;
3. Level 2: fully collaborative BIM approach.

In the proposed BeF, it is also believed that the transformation from traditional to BIM based practice requires a progressive transformation, therefore, this framework also presents how incrementally traditional practice could be transferred into BIM (Miettinen and Paavola, 2014). Each criterion has been divided aligning with these three levels according to their own character: column F presents non-BIM was involved during the practice; column G to J present a progress improvement of BIM used in practice till collaborative BIM. Different category in column 'Maximum level' indicates the maximum scale that a criterion could meet, where 'Current Level' and 'Target Level' identify project's current and target BIM usage level, this could allow the application 1 as shown in Figure 6-1. In other cases, they can be replaced by project 'A', 'B' and 'C' for comparison between projects.

As shown in Figure 6-1, User will simply input their appropriate maturity level under 'User fill in' column, their weight can be automatically calculated and the summation can used for comparison.

For example, the full point for criterion 'Milestone and delivery' is 0.48: capability level 0 will receive 0 point, level 1 will receive 0.24 points and level 2 will receive 0.48. By selecting level '1' or level '2' in 4P and 4S in the Excel table, 0.24 or 0.48 will be obtained to represent the capability of BIM for this particular criterion.

With such self-assessment framework, BIM user could identify their current performance for continuously improvement through an easier, efficient and consistent approach (Xu and Yang, 2003). In addition, the steps toward collaborative BIM have been elaborated from column F to J to guide user to move for the next level of BIM using.

The final credit gained will be accumulated both at the end of each dimension and the whole assessment framework. This allows the result to be reviewed by different beneficiaries, such as the management team, project team and stakeholders etc. (Du et al., 2014) to point out existing weakness for an overall improvement.

1-1 BIM Deployment

Level of BIM capabilities - 1. Project management

							Maximum level		Current level			Target level		
Factor (selected for China Context only)	Description	0	1	2	3	4	Proficiency level	Maximum weight available (Based on context)	Proficiency level (User fill in)	Sub-factor weight	Factor weight	Proficiency level (User fill in)	Sub-factor weight	Factor weight
Administration deployment														
Milestone and deliverable	Project delivery schedule required to be clearly defined and followed by all project parties e.g. exact time and data format need to be followed	Project delivery information are briefly mentioned, only date has been mentioned	Project delivery information are clearly mentioned with a format & date requirement	Project delivery clearly defined & executed throughout the entire project with mandatory enforcement by manager			2	0.480	1	0.240	1.040	2	0.480	1.880
LOD	Level of Detail describes the amount of information delivered by the model in order to fit the purpose of a specific drawing	No LOD has been defined or used	LOD defined but limited to certain project stages	LOD has been extensively utilised throughout the entire project	LOD has been aligned with industry standards e.g. MVD & BIM and additional customised requirement		3	0.480	1	0.160		2	0.320	
Naming and folder structure	A well organised and followed naming & folder convention nomenclature could improve working efficiency	No naming convention has been adopted	Naming convention has been adopted for certain area of the design work and business documentation	Naming convention utilised throughout the entire project, for all business documentation, by all parties			2	0.240 / 0.480 (1.440)	1	0.120		2	0.240	
Technical deployment														
Information interoperability (e.g. IFC)	An open standard for compatible collaboration environment is the pre-requisite to adopt BIM and it encourages the integrated design process among all stakeholders	No folder structure has been adopted	Folder structure has been adapted by part of the project teams	Full interoperability was achieved by machine to machine linkage or limited use of IFC	Full interoperability has been achieved by using IFC throughout the project lifecycle		2	0.240 (2.400)	2	0.240		2	0.240	
		No compatibility between platform or teams was considered	Limited interoperability has been achieved by adopting tools from the same vendor e.g. Revit, Bentley or the use of IFC				3	0.480 (0.960)	1	0.160		3	0.480	
Technical reuse	Essential skills that required for BIM activities e.g. modelling skill, coordination skills, etc.	No relevant BIM technology has been used	Basic semantic modelling skill e.g. Revit, Bentley etc.	Advanced BIM usage skills, collaboration related activities	BIM skills for construction purposes e.g. Quantity takeoff	Object scheduling e.g. schedule embedded within the BIM model, IFC includes info of quantities & materials	4	0.480	2	0.120		1	0.120	
								Total:			**1.040**		**Total:**	**1.880**

Figure 6-1 Example of proposed BIM evaluation Framework (BeF)

6.2 Validation result & analysis

The developed BIM implementation and assessment framework is intended to be used by BIM management personnel, hence the evaluator used in this research is ASZ BIM manager since he is the only person who can provide the most objective assessment result.

6.2.1 Comparison among practical projects

Projects A-E result analysis

There are five projects were selected for the evaluation stage (Table 6-1). The evaluation was conducted at the end of 2014, projects (C, D and E) completed in early 2014 and late 2013 have been defined as 'recent completed' project. Projects A, B and C were complete by the same team. Based on the hypothesis made before, their BIM maturity and performance should be continuously improving along with time.

Projects	Team	Design completed time	Expected result	Actual BIM level	nD BIM Model quality	Actual project performance	Project character
A	1	2009	Lowest	7.644%	3.14	46	Small scale project
B	1	2012	Low	10.894%	3.43	48	Small scale project
C	1	Early 2014	High	31.954%	5.86	66	Landmark, large scale project, high requirement for BIM performance from client
D	2	Late 2013	High	15.44%	4.57	54	Tight budget and time, limited human resource
E	3	Late 2013	High	41.581%	6.57	64	New management system

Table 6-1 Validation result of project A-E

Project A was designed in 2009 for a simple office building in Shen Zhen. As one of the earliest BIM based project, its project outcome and

BIM usage level are the lowest in all five projects as expected in the framework.

Project B was designed in 2012 for an ultra-high rise finance center, including hotel, luxury flat, shopping centers and building office in Shen Zhen, its project outcome and BIM maturity is higher than project A.

Project C was designed in early 2014 for a mixed shopping mall, office building, hotel etc. in Cheng Du. Its BIM usage level and project outcome is much higher than project A & B.

Project D was designed in late 2013 by team 2 for an R & D type of office building. As one of the recent completed project, its BIM usage level and project outcome should be similar to project C, but it is much lower although it is still higher than Project B.

Project E was designed in late 2013 as well for super high rise office building, its BIM usage level and project outcome have received the highest among all five projects. All these comparisons demonstrated that the test results from the developed framework aligns well with the expected outcomes.

<u>nD BIM model comparison</u>

In order to reveal how well the criteria included in the BeF have been carried out during the actually design process, hence the comprehensiveness and operation of the nD BIM model of selected five projects have been analyzed.

The result presented in Table 6-2 overleaf shows a similar trend to the BIM usage level assessment that achieved in each project, this proved the maturity from BIM manager's perspective also align with individual's BIM proficiency.

Items	A	B	C	D	E
LOD (lifecycle adoption)	4	4	6	4	7
Naming/folder structure	2	4	6	5	7
Compatibility	6	6	8	6	6
Information richness	3	3	5	6	4
4D BIM	2	2	6	2	6
Data access/modification right	3	3	6	4	8
Customized template/library component	2	2	4	5	8
Average	3.14	3.43	5.86	4.57	6.57
1-10: 1 = poor; 10 = good					

Table 6-2 nD BIM model comparison among project A-E

Project performance comparison

In order to compare the BIM usage level and project performance in the same project, a list of Key Performance Indicators (KPIs) have been selected from literature, discussed and approved by Arup management team to represent the project's performance in four aspects: profit, quality, efficiency and client's satisfaction, as shown in Table 6-3 overleaf.

The selection of KPIs should follow several rules: clear, measurable, and relevant to the measure object etc. However, some of the above-mentioned KPIs are still not practical in reality.

-KPIs that are relevant to profit are hard to track or link to BIM. Since there is no record to clarify the exact amount of cost for BIM hardware, software expenditures and training etc. as those expenses are one time capital cost which can have influence in more than one project.

-Return on investment (ROI) cannot be obtained due to commercial sensitivity. The calculation of ROI is also complex, it involves land value, various regulatory processes approvals, and other costs.

-The definition for rework is hard to separate from normal tasks, as this was not recorded during practical design work. For rework during construction stage, it will only be managed by contractor.

-The efficiency and what has been saved or improved by BIM during the design or construction stage is hard to be quantified. Based on the quality, experience, maturity of the ability of the team and management approach, there will be a downside at the initial implementation stage, however, in long term, BIM could help to achieve a better planning during design stage, with a better coordination performance.

Category	Items	Formula	Replacement	Sources	Availability
Profit (of project: design work in this case)	Profitability (BIM)	Profit/revenue%	1-10 scale	(ONS, 2010)	Yes
	Profitability (Overall)	Profit/revenue%	1-10 scale		Yes
	BIM ROI (coordination/cost estimation)	BIM net saving/BIM cost%	Total save-total cost (1-10)	(Giel and Issa, 2013)	Yes
Quality (Ali et al., 2013) (of the developed nD BIM model and final product delivery to the client)	Model value to cost-estimation	1-10 scale	-		Yes
	Rework	Rework cost/total cost%	-	(Ali et al., 2013)	NA
	Quality management system	1-10 scale	-	(Luu et al., 2008)	Yes
	Contract & legal disputes	1-10 scale	-	(Ling et al., 2009)	Yes
	Environment & sustainability	1-10 scale	-	(Najafabadi, 2013)	Yes
	Product defects	1-10 scale	-	(ONS, 2010)	Yes
Efficiency (work done in unit time)	Efficiency ratio	Cost/revenue%	-	(Yu et al., 2007)	NA
	Design/construction cost ratio	(Actual-estimated cost)/estimated %	-	(Luu et al., 2008)	NA
	Design/construction schedule ratio	(Actual-estimated time)/estimated %	-		NA
Client's satisfaction	Service satisfaction	1-10 scale	-	(Luu et al., 2008, ONS, 2010)	Yes
	Product satisfaction	1-10 scale	-		Yes
	Customer's repeatability	1-10 scale	-	(Ali et al., 2013)	NA
	Innovation	1-10 scale	-	(Najafabadi, 2013)	Yes
1-10: 1 – poor; 10 – good					

Table 6-3 Key Performance Indicators for project A-E

The selection of KPIs should follow several rules: clear, measurable, and relevant to the measure object etc. However, some of the above-mentioned KPIs are still not practical in reality.

-KPIs that are relevant to profit are hard to track or link to BIM. Since there is no record to clarify the exact amount of cost for BIM hardware, software expenditures and training etc. as those expenses are one time capital cost which can have influence in more than one project.

-Return on investment (ROI) cannot be obtained due to commercial sensitivity. The calculation of ROI is also complex, it involves land value, various regulatory processes approvals, and other costs.

-The definition for rework is hard to separate from normal tasks, as this was not recorded during practical design work. For rework during construction stage, it will only be managed by contractor.

-The efficiency and what has been saved or improved by BIM during the design or construction stage is hard to be quantified. Based on the quality, experience, maturity of the ability of the team and management approach, there will be a downside at the initial implementation stage, however, in long term, BIM could help to achieve a better planning during design stage, with a better coordination performance.

The 0 - 10 rating scales were adopted where 0 presents BIM has no benefit in the adoption of this particular aspect, and 10 presents an extremely satisfied level. Table 6-4 demonstrates the result collected from BIM manager in ASZ.

	Items	A	B	C	D	E
Designer perspective	Profitability (BIM)	1	1	2	2	1
	Profitability (Overall)	1	1	3	3	5
	BIM ROI (Coordination)	0	0	2	3	1
	BIM ROI (Cost estimation)	1	1	5	3	1
	Model value to cost-estimation (Architecture)	1	1	1	6	6
	Model value to cost-estimation (Structure)	1	1	6	7	8
	Model value to cost-estimation (MEP)	1	1	1	7	8
	Model value to cost-estimation (4D scheduling)	1	1	5	4	4
	Quality management system	1	1	6	6	6
	Contract & legal disputes	1	1	7	7	8
	Environment & sustainability	1	1	1	0	4
Client perspective	Product defects	9	9	7	8	7
	Service satisfaction	8	9	8	8	8
	Product satisfaction	10	10	9	8	8
	Innovation	6	5	8	6	8
1-10: 1 = poor; 10 = good						
Total		43	43	71	78	83

Table 6-4 Project performance comparison for project A-E

Based on the hypothesis, project A & B should have a lower satisfaction in all aspects compared to project C, D & E. however, this is not the case in some of the KPIs as shown in Table 6-4 above. Project A & B are simple structure buildings, which is easier to be delivered and meet customer's requirement for a higher satisfaction. Moreover, during the early adoption stage of BIM in China's construction industry, the use of BIM in project A & B has just started and under an exploration stage. The expected BIM achievement and functions in business activities were simple and easier to be achieved (for example, the main objective is to generate 3D building model). However, in a later stage, when the project C, D and E was conducted, client's knowledge of BIM has improved according to public reports from more developed countries such as US e.g. (McGraw-Hill construction, 2008, McGraw-Hill Construction, 2009a, McGrw_Hill Construction, 2012). But due to the limited development of BIM in China, client always expect more from the designer and contractor, but which could not meet their expectations. Therefore a relatively low satisfaction level has been achieved.

The total value achieved in Table 6-4 shows there is a clear improvement in term of project performance in its all aspects (profit, quality, efficiency and customer satisfaction) with the progress made in BIM in reality.

Factors that affect BIM usage level

Based on the BIM usage level evaluated by the proposed BeF, discussion was conducted with BIM manager in ASZ, reasons that could affect BIM usage level have been concluded.

In the current China AEC industry, the traditional project management approach, people's professional cultures and working environment are very difficult to be changed towards BIM compliance, which have been believed as one of the main challenges for using BIM. What's more, there is no simple, quick and effective solution for this. Instead of replace the traditional approach completely, it is better to embed BIM concept into the existing method and process, therefore to facilitate further development within the local context. However, this causes majority of the project are still following a traditional 2D CAD based approach, while BIM is only used for specific purposes, e.g. 3D BIM models are created for visualization and clash detection only, information within the BIM model has not been reused by other stages; The contractors still follow 2D based construction drawings, while the BIM model received from designer are mainly used for 4D based scheduling to support construction management approach.

At the beginning years of BIM adoption, technology was the main focus, hence project A & B were focusing on 3D modelling and coordination among disciplines (Brynjolfsson, 1993, Jung and Gibson, 1999). Other aspects of BIM e.g. management, organizational & stakeholders involvement etc. was not considered in those projects which therefore led to a lower BIM usage level (5.9% & 9.2% respectively); project C is an iconic large-scale mixed-use projects with multiple functions. As one of the most recent projects, industry's awareness on BIM has been largely expanded. The client had a higher requirement on the multiple aspects of BIM adoption, such as project management aspect, coordination and collaboration among stakeholders, therefore a higher BIM maturity level (30.3%) was achieved.

Management paradigm

The adoption of BIM and its usage level is also based on the project manager's individual method, as well as available resources. Project D has a tight project schedule and labor resource, where only one engineer was working on a single BIM model. Moreover, the project manager's own management paradigm stayed more with traditional approach. Therefore, only technological aspect of BIM was adopted (e.g. 3D visualization, clash detection and 4D BIM etc.), which is similar to project A & B. However, those criteria have not been particularly emphasized in project D, such as *BIM deployment* and industry's awareness (*stakeholder involvement*), still received a higher BIM level compared to earlier project. Therefore, compared to project D, a lower level of BIM level was achieved (14.773%) but still higher than earlier projects A & B.

Starting from project E, a new systematic method of BIM management was adopted and a much higher maturity was achieved (39.998%). The method includes: development of standardized template, standards, continuously improved repository and operation process etc. It largely reduced the requirement on individual's BIM competency and improve the quality of collaboration between different parties (e.g. visualize all building's information). From the organizational level, their BIM implementation has standardized the BIM adoption procedures in its project hence the organization's management maturity has been increased.

Contractual and legal aspect

The BIM related profit issues among disciplines have not been clearly defined in the current contract/legal practice, which is another barrier for BIM adoption throughout building lifecycle. In the case of ASZ, the client normally has a contract relationship with designer and contractor individually, however, there is no contract relationship between designer and contractor. Therefore without a contract framework (e.g. (ConsensusDocs, 2011)), dispute and potential risk could happen especially when there is a need for collaboration and coordination with other partners in BIM implementation.

Therefore, the selection of BIM function, or the priority of key implementation area of BIM is largely based on the lifecycle operation of the project, management requirement, project's character, client's need etc. Based on the vertical and horizontal comparison among those five projects, it proved the proposed BeF has a good correlation to the final project outcomes, e.g. where a higher level of evaluation result suggests a better project outcome. In another word, adopting this framework can improve the BIM implementation level and ultimately project performance.

6.2.2 Comparison with existing methods

In order to validate the accuracy of the BeF, existing assessment tools (e.g. I-CMM, BPM, OBIM & ABMF in Chapter 2) have also been used by the BIM manager to evaluate all five projects in previous step. As shown in Table 6-5 below, results obtained from all five applied assessment methods provide a similar trend. As discussed, existing assessment methods focus more on a single aspect of BIM implementation, so by calculating their average, it could provide a more comprehensive result considering all aspects. The comparison between their average values and the result obtained by BeF are shown in Table 6-5 and Figure 6-2.

All result shows the proposed BeF is effective and efficient and meet user satisfaction.

Project	Result					
	Proposed tool	Other existing tools				
		I-CMM	BPM	OBIM	ABMF	Average
A	7.644%	15.5%	34.4%	4.4%	27%	19.6%
B	10.894%	23.7%	37.5%	6.7%	27%	23.7%
C	31.954%	44.8%	62.5%	11.1%	40%	39.6%
D	15.44%	39.5%	31.3%	13.3%	33%	29.3%
E	41.581%	35.7%	43.8%	15.6%	35%	32.5%

Table 6-5 Project A-E results comparison between existing methods' average value and BeF

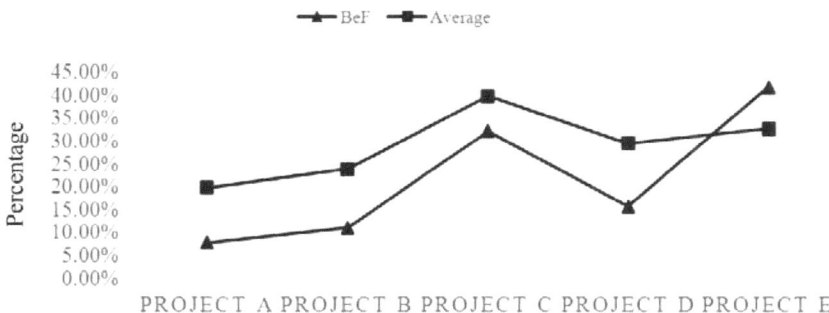

Figure 6-2 Project A-E results comparison between existing methods' average value and BeF

6.2.3 User's feedback

It takes about 30 minutes for the BIM manager from ASZ to fully understand and digest the concept, definition of criteria proposed, the assessment framework's logic structure and intention. The developed BIM evaluation framework covers major BIM implementation areas during the design stage. Moreover, BeF has fully considered all stakeholders' involvement, this is the very key for BIM implementation in most cases. The user's feedback is concluded as below:

1. The proposed framework can effectively assist project manager for BIM strategy planning: business activities can be conducted according to their priorities, wise decision can be made when there is a resource constraint for a higher level of BIM implementation;
2. BeF illustrates all important implementation criteria to all stakeholders, the evaluation process was conducted under a transparent circumstance, and helped to facilitate comparison and improvement.
3. Proposed BeF has displayed all criteria effectively in a structured and systematic way which can be used to evaluate and monitor business improvements (BIMtaskgroup, 2012);
4. The framework can be used to evaluate whether or not the organization is ready or qualified for BIM based project, especially on how good or bad for their BIM implementation capability.
5. The proposed framework can also be applied for education/training purpose, to give a general background of BIM concept to the trainees, as well as their importance to the various objectives. Based on the weight of each criteria, training and education plans can be produced to improve the operational proficiency, BIM usage level and project performance.
6. The assessment framework is based on Microsoft Excel, it is easy to use, without additional training or hardware requirement, or constrained by internet or complex installation process;
7. The assessment process can be performed by any individual employee or a management team;

8. The proposed assessment framework can be easily modified and adapted for changed context, e.g. add or remove criteria, change of weightage system or add or remove future objectives for specific scenario (Du et al., 2014, Wang et al., 2005).

6.3 Sensitivity analysis

6.3.1 Ranking score of focus schemes

As included in the AHP questionnaire, the second section compared the impact of each focus scheme towards all criteria within the BeF. There are sixty five criteria in total, which generates sixty five questions and matrices. The weight of each focus scheme can be obtained by calculating the eigenvector of each matrix. In total, sixty five set of weight for each focus scheme can be generated. In order to obtain the ranking score of each focus scheme in ASZ, the composite weight was used. This can be contained by using the weight of each focus scheme multiplies the global priority of each corresponding criteria, and add with other sixty four pairs.

Alternatives	Weight				Total (%)
	Project management	Organizational management	BIM application in lifecycle	Partner involvement	
Commercial value	16.22	29.64	5.16	8.11	59.13
Customer satisfaction	6.82	13.35	2.24	3.99	26.40
Sustainability	3.69	7.65	0.99	2.15	14.48
				Total:	100

Table 6-6 Ranking order of future objectives
& contribution from each dimension

In a similar fashion, the ranking order of future objectives (Table 6-6) was also obtained from MIR, the composite value shows ASZ has an emphasis on the creation of commercial value (59.13%) outstripping the other two. Also there are less resources allocated on other alternatives, especially for development in sustainability (14.48%). Thus the result shows there is a stable emphasis in commercial value for their current development (Zhu et al., 2005).

6.3.2 New priorities & strategy

Besides commercial profits, ASZ also needs to consider the need from local communities, the development for human society and global competitiveness, and more sustainable and user oriented products and services. Hence it is necessary for ASZ to shift their existing priority to have more emphasis on sustainability. Dimensions and factors contained within the proposed hierarchy model are interconnected and closely related to each other. In addition, due to the summation of each criterion's weight (in percentage) is equal to 1 (100%), improvement of any factor's weightage could result in reducing others'. Hence it is necessary to reveal the impact of each criteria to each focus scheme: either there is a positive impact or negative impact, which can be improved or reduced accordingly.

Based on the AHP questionnaire collected from ASZ, MIR has calculated the relationship between a specific criteria and focus schemes. For example, MIR obtained an influencing chart for criterion: *energy & sustainability* (E&S) regarding three alternatives:

Energy & sustainability (%)				
Alternatives	Minimum 0	Current 6.1	Maximum 100	Impact
Customer's satisfaction	26.39	26.40	26.50	0
Sustainability	14.34	14.48	16.61	0.0227
Commercial value	59.27	59.13	56.88	-0.0227

Table 6-7 Example of sensitivity analysis: impact of *Sustainability* towards Focus schemes

As provided from Table 6-7, the current global priority of E&S is 6.1%, by improving its global weight, there will be a 0.0227% improvement for every 1% improving. The tendency are generated from MIR as Figure 6-3 below:

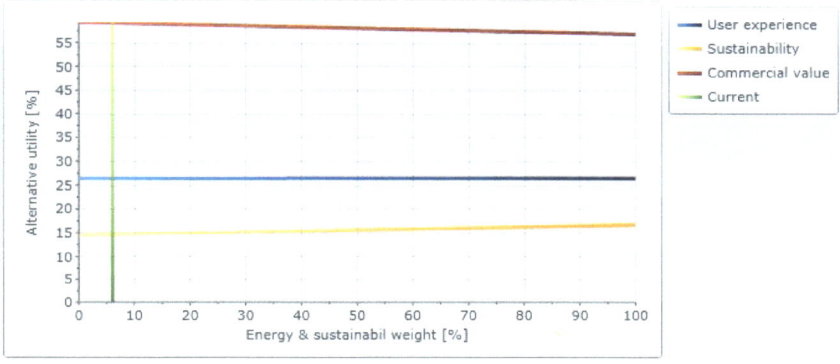

Figure 6-3 Example of sensitivity analysis: impact of *Sustainability* towards Focus schemes

As shown in Table 6-8, a positive value means an incremental trend, while a negative value means a reduction trend to that focus scheme. The perturbation to the original options' ranking has been adjusted by the researcher together with BIM manager by altering the priorities of the criteria within BeF. Its final decision is shown in Table 6-8: under column 'Adjustment trend'. The adjustment decision has been shown. Please note, two decimal places have remained.

Dimensions & factors to be adjusted	Influence factor		
	Commercial value	Sustainability	Customer's satisfaction
Project management	**0.02**	**-0.01**	**-0.01**
Technical & tools requirement	-0.01	0	0.01
BIM deployment	0	0	0
Legal & contractual	-0.03	0.03	0
Log management	-0.02	0.02	0
Project performance evaluation	0.02	-0.01	0
Customer & quality assurance	0	-0.01	0.01
Resource allocation	0.01	0	0
Application management	**0.03**	**-0.03**	**0**
Energy & sustainability	-0.02	0.02	0
Quantity take-off	0	0	0
Organization management	**-0.01**	**0.01**	**0**
BIM performance evaluation	-0.04	0.02	0.02
Consistent vision	-0.08	0.04	0.04
staffing skill, plan & training	0.04	-0.02	-0.02
Business reengineering	0.06	-0.02	-0.03
Stakeholder involvement	**-0.03**	**0.01**	**0.02**
Government & public involvement	-0.02	0.02	0
Software vendor	0.01	-0.01	0
Manufacture	0.01	-0.01	0

Table 6-8 Sensitivity analysis of factor & dimension change trend

By shifting the priority of the factors in Table 6-8 above, it will affect all other criteria. The new set of priority are shown in Table 6-9 below:

Hierarchy Level	Goal/Dimension/Criteria/Sub-criteria	Original weight (%)	New global weight (%)
1	Goal: Future objective orientation of modern AEC organization, a case study of ASZ	100	100
2	**Project management**	26.7	17.8
3	Project execution	9.9	6.6
4	BIM deployment	2.4	0.7
4	Roles & Responsibilities	0.9	0.3
4	Collaboration Environment	1.4	0.4
4	Process & Plan of work	1.6	0.5
4	Change order	1.3	0.4
4	Technical & tools requirement	1.1	2
4	Innovation	1.2	0.4
3	Project maintenance/ administration	16.8	11.2
4	Legal & contractual	2.3	3.7
4	Resource allocation	2.2	0.4
4	Log management	1.2	5.3
4	Customer & quality assurance	3.1	0.5
4	Data management	1.6	0.3
4	Project performance evaluation	3.1	0.5
4	Risk control	3.3	0.5
2	**Application management**	8.4	24
3	3D visualization, coordination & clash detection	0.7	0.4
3	Quantity take-off	1.5	0.9
3	Energy & sustainability	0.5	3.4
3	4D schedule management	2	1.2
3	5D budget control	2.5	1.5
3	FM	0.7	0.5

Hierarchy Level	Goal/Dimension/Criteria/Sub-criteria	Original weight (%)	New global weight (%)
3	Safety & environment	0.5	0.3
2	**Organizational dimension**	50.6	33.7
3	Consistent vision	3.4	17.8
3	BIM performance evaluation	25.3	9.5
3	Business process reengineering	12.3	2.7
3	Staffing skill, plan & training	6.3	2.4
3	Internal R&D	3.4	1.3
2	**Stakeholder involvement**	14.2	24.5
3	Owner/client	5.1	2.3
3	Designer	2	0.9
3	Contractor	1.2	0.5
3	Manufacturer	1	0.8
3	Software vendor	0.7	0.3
3	Institution	0.7	0.3
3	Government & public	3.4	19.6

Table 6-9 Global priority before and after Sensitivity analysis

A new option ranking was also obtained as shown in Figure 6-4, where the percentage of Commercial value was reduced to 48.11%, customer satisfaction only improved by about 2%, and the percentage of sustainability was improved by about 11%. The adjustment did not provide a very obvious difference compared to the original trend. Since at this initial transformation stage, the strategy and action taken should be more conservative, to avoid any potential profit lose and risks. As such, a better decision and strategy can be made by BIM manager (Triantaphyllou, 1997).

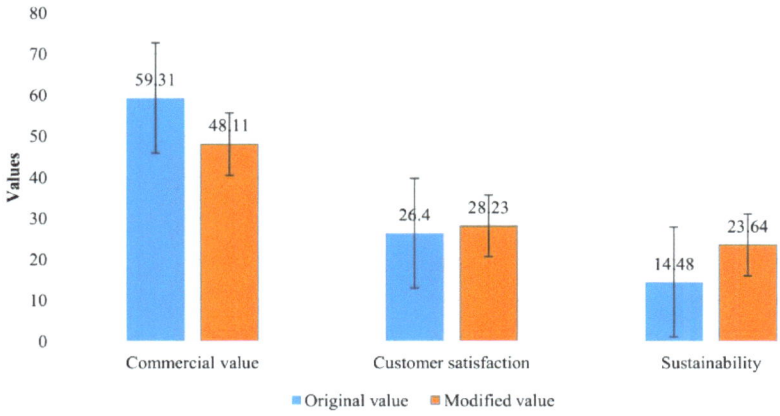

Figure 6-4 Comparison of alternatives rankings before and after factor priorities shifting

6.4 Summary

The criteria concluded from Delphi method was renamed, merged, removed and re-categorized to keep only managerial criteria to form a hierarchy AHP model that was approved by ASZ. In order to obtain the priority of each criteria, as well as current ASZ's focus schemes, Analytical Hierarchy Process (AHP) method was adopted. The global and local weights of each criteria was calculated by using the commercial application tools: Make It Rational (MIR). The result shows ASZ has a focus on the BIM adoption from the organization level.

Based on the weightage system, a BIM evaluation Framework (BeF) has been developed. Firstly, a link among all managerial criteria and technological criteria was established. Secondly, each criteria was divided into various levels, from non-BIM capability to collaborative BIM capability. Thus, the third research questions has been addressed: *'How to develop a multi-criteria decision making tool to assist in strategic BIM implementation and assessment for an organization?'*

In order to validate the efficiency, effectiveness and user satisfaction (e.g. BIM manager) of the developed evaluation framework, BeF was used to assess five completed projects from horizontal and vertical angles. The validation shows a good correlation - if an AEC project received a higher credit by using BeF, it means a better project performance in terms of profit, quality, efficiency and client satisfaction. The results also revealed that the level of BIM usage could also be related to the client's requirement, management paradigm and contract & legal aspects.

In order to validate the accuracy of the proposed BeF, other existing BIM assessment frameworks (provided by: I-CMM, BMP, OBIM and ABMF) have been selected to conduct the validation first. The results obtained by each individual assessment method all have a similar trend compared to the result obtained by BeF.

There are three focus schemes proposed by ASZ management team. The current ranking score obtained by MIR shows ASZ's attitude is focusing on the creation of commercial value (59.31%). In order to meet the need of company's new objective on sustainability and customer's satisfaction, sensitivity analysis was adopted. The influence of each criteria towards each focus scheme has been identified by MIR, the weight of criteria with a positive influence to *sustainability* will be improved, and criteria with a positive influence to *commercial value* will be reduced. After shifting the criteria's priority, the new rankings for ASZ's focus schemes are: commercial value (48.11%), sustainability (23.64%) and customer's satisfaction (28.23%).

7 CONCLUSION

In traditional BIM based projects, the implementation approach is based on the users' experience and preference. As a result, the overall BIM capability and outcome is unexpected and uncontrollable. With the progress made in the BIM field, its advantages in different areas have been highlighted by the users, for example: sustainability, budget and schedule control. Therefore, how BIM could influence each specific object becomes critical. Especially how to achieve the maximum outcome through minimum input. Such a framework could also serve as a new way of managing BIM in a more strategic manner: to assess before adopt and to assess during adoption for better performance.

The foundation of this framework is based on existing BIM documentations, where the author has integrated most existing BIM standards, guidelines, protocols and existing BIM assessment tools etc. to ensure the comprehensive of this framework, generic and applicable to all cases, regardless its types of project, cultures of stakeholders or even political environment. In the next stage, the proposed framework was tested in a case study and a real project for its reliability and usability in reality.

The use of BIM could be influenced by variables like policies, cultures, business structures and legislation etc. Hence the development of BIM strategy was customized based on a specific region: China. A decision making framework from the management's perspective has been abstracted from the revised BiF in previous step. The Analytical Hierarchy Process (AHP) method has been employed to analyze the weighting system in a design consultant: ASZ. Wherein the final BIM evaluation Framework (BeF) was developed with the following applications: to assist project manager or BIM manager to develop

BIM strategic planning in practical adoption based on the priority of each criterion; to assess the current BIM usage level, identify the weakness, strength and what is missing from the current practice; to be adopted by client for prequalification selection for the optimized designer during the bidding stage.

The ranking order of three focus schemes proposed by ASZ was obtained: it shows the current attitude of ASZ is on 'commercial value' (59.31%). In order to have more focus on the 'sustainability' as well as 'customer's satisfaction', sensitivity analysis was adopted. By modifying the priorities of criteria within the hierarchy model, people's attitude on different focus scheme can be shifted as well.

Research contribution

The first contribution of this research is to the BIM implementation strategies, in the AEC industry context: a generic BIM implementation Framework (BiF) which aims to implement BIM from five dimensions: project, data, application and organizational management as well as stakeholder involvement was proposed. The BiF will not be constrained by culture, policy, political environment, discipline and project types etc.

Secondly, this research has contribution towards the AEC industry of China. The BiF was refined by Delphi in China, which can be used by other organizations in China. Moreover, new focus scheme can be added to the original AHP framework proposed by ASZ. In addition, they Delphi and AHP approach can be replicated for any other decision making problem.

Thirdly, the proposed BeF has contributed to the BIM adoption within ASZ: to facilitate ASZ to improve their level of BIM usage by revealing the weak point and missing area, to apply the highest prioritized criteria to meet a higher BIM usage level if the resource is limited.

Lastly, the developed BeF has also contributed to the assessment of BIM usage level, since less attention has been paid to the involvement of stakeholders during the design stage, instead, existing assessments have a focus mainly on the data process. Even though some existing method have include BIM evaluation from the organizational level, while the assessment criteria still require rigorous selection and validation based on specific context, for example countries. The validation stage has proved the internal variables (e.g. project types, client requirement, project management approach and budget limit etc.) could influence the BIM usage level in a micro scale, while the external variables' impact (e.g. culture and political environment etc.) still require further test.

Most importantly, the weighting system should be calculated in a reasonably way, instead of using average value or assumption. In addition, organization's focus schemes should be considered therefore to provide a more appropriate strategic planning for BIM development.

Recommendation for future work

The author is currently working in Building and Construction Authority in Singapore. Singapore as one of the advanced BIM pioneers in the world has accomplished in many aspects in the game changing technology (e.g. BIM, VDC etc.) to improve its productivity. From the author's perspective, relevant organizations should start to focus on the assessment of individual / teams' capability in term of BIM performance and capability, hence to qualify project participants for a better outcome. Meaning while, a more informed strategy can be developed to fully use available resources.

REFERENCES

ADRIAANSE, A., VOORDIJK, H. & DEWULF, G. 2010. The use of interorganizational ICT in United States construction projects. *Automation in Construction,* 19, 73-83.

AEC (UK) INITIATIVE 2012. AEC (UK) BIM Protocol - Implementing UK BIM Standards for the Architectural, Engineering and Construction industry. *Updated to unify protocols outlined in AEC (UK) BIM Standard for Revit and Bentley Building.* UK: AEC (UK) Initiative.

AIA 2007. Integrated Project Delivery: A Guide. The American Institute of Architects.

AIA 2013b. AIA Document E203TM – Building Information Modelling and Digital Data Exhibit.

AISH, R. Three-dimensional Input and Visualization. Computer-Aided Architectural Design Futures: International Conference on Computer-Aided Architectural Design, 1986 London: Butterworths.

ALI, H. A. E. M., AL-SULAIHI, I. A. & AL-GAHTANI, K. S. 2013. Indicators for measuring performance of building construction companies in Kingdom of Saudi Arabia. *Journal of King Saud University - Engineering Sciences,* 25, 125-134.

ALSHAWI, M., GOULDING, J., KHOSROWSHAHI, F., LOU, E. & UNDERWOOD, J. 2008. Strategic positioning of IT in construction, an industry leaders. *proceedings of the International Conference on Computing in Civil and Building Engineering.* Nottingham University Press.

ALYAMI, S. 2015. *The Development of Sustainable Assessment Method for Saudi Arabia Built Environment.* Ph.D, Cardiff University.

ARANDA-MENA, G., CRAWFORD, J., CHEVEZ, A. & FROESE, T. 2009. Building information modelling demystified: does it make business sense to adopt BIM? *International Journal of Managing Projects in Business,* 2, 419-434.

ARAYICI, Y. & COATES, P. 2013. Operational Knowledge for BIM Adoption and Implementation for Lean Efficiency Gains. *Journal of Entrepreneurship and Innovation Management,* 1, 1-20.

ARAYICI, Y., COATES, P., KOSKELA, L., KAGIOGLOU, M., USHER, C. & O'REILLY, K. 2011. Technology adoption in the BIM implementation for lean architectural practice. *Automation in Construction,* 20, 189-195.

ARAYICI, Y., KHOSROWSHAHI, Y., PONTING, A. M. & MIHINDU, S. 2009. Towards implementation of building information modelling in the construction industry. *5th International Conference on Construction in the 21st Century (CITC-V) "Collaboration and Integration in Engineering, Management and Technology".* Istanbul, Turkey.

ARUP. 2015. *Transforming the urban landscape* [Online]. Available: http://www.arup.com/ Global_locations/Greater_China.aspx?sc_lang=en-GB [Accessed 10th September 2015].

BARLISH, K. & SULLIVAN, K. 2012. How to measure the benefits of BIM — A case study approach. *Automation in Construction*, 24, 149-159.

BARRY, G., DANIEL, H. & PAULA, B. 2012. Does size matter? Experiences and perspectives of BIM implementation from large and SME construction contractors. *1st UK Academic Conference on Building Information Management (BIM) 2012.* 5-7 September 2012, Northumbria University, Newcastle upon Tyne, UK.

BECERIK-GERBER, B., JAZIZADEH, F., LI, N. & CALIS, G. 2012. Application Areas and Data Requirements for BIM-Enabled Facilities Management. *J. Constr. Eng. Manage*, 138, 431-442.

BENDER, A. D., STRACK, A. E., EBRIGHT, G. W. & HAUNALTER, G. V. 1969. Delphic study examines developments in medicine. *Futures*

BIMIWG 2011. BIM Working Party Strategy Paper, A report for the Government Construction Client Group. BIM Industry Working Group.

BIMTASKGROUP 2012. Investors Report Building Information Modelling (BIM).

BLOOM, N. & REENEN, J. V. 2006. Measuring and Explaining Management Practices Across Firms and Countries. *NBER Working Paper No. 12216.* the National Bureau of Economic Research.

BMCUP 2013. Building Information Modeling Design Standard for Civil Building (in Chinese). Beijing: Beijing Municipal Commission of Urban Planning, Beijing Exploration and Design Association.

BOZBURA, F., BESKESE, A. & KAHRAMAN, C. 2007. Prioritization of human capital measurement indicators using fuzzy AHP. *Expert Systems with Applications*, 32, 1100-1112.

BRYNJOLFSSON, E. 1993. The productivity paradox of information technology. *Communications of the ACM*, 36, 66-77.

BSI 2013. PAS 1192-2:2013 Specification for information management for the capital/delivery phase of construction proejcts using building information modelling. United Kingdom: British Standards Institution.

BSI 2014a. BS 1192-4:2014 Collaborative production of informationPart 4: Fulfilling employer's information exchange requirements using COBie – Code of practice. London British Standard Institution.

BSI 2014b. PAS 1192-3:2014 Specification for information management for the operational phase of assets using building information modelling UK: British Standard Institution.

CABINET OFFICE 2011. Government Construction Strategy. UK.

CAMP, R. C. 1995. *Business Process Benchmarking: Finding and Implementing Best Practices*, Milwaukee, WI., ASQC Quality Press.

CHANG, C.-W., WU, C.-R., LIN, C.-T. & CHEN, H.-C. 2007a. An application of AHP and sensitivity analysis for selecting the best slicing machine. *Computers & Industrial Engineering,* 52**,** 296-307.

CHANG, K.-F., CHIANG, C.-M. & CHOU, P.-C. 2007b. Adapting aspects of GBTool 2005—searching for suitability in Taiwan. *Building and Environment,* 42**,** 310-316.

CHEN, K. & LI, H. 2014. A BIM management model using Delphi approach. *European Group for Intelligent Computing in Engineering.* Cardiff, The United Kingdom.

CHEN, Y., DIB, H. & COX, R. F. 2014. A measurement model of building information modelling maturity. *Construction Innovation* 14.

CHIEN, K.-F., WU, Z.-H. & HUANG, S.-C. 2014. Identifying and assessing critical risk factors for BIM projects: Empirical study. *Automation in Construction,* 45**,** 1-15.

CHURCHER, D. & RICHARDS, M. 2013. Cross-discipline design deliverables for BIM Phase 1 report - Strategy Document. British

CIOB. 2015. *BIM around the world - country by country* [Online]. Available: http://www.construction-manager.co.uk/agenda/bim-around-world-country-country/ [Accessed 20th June 2015].

COLE, R. J. 2005. Building environmental assessment methods: redefining intentions and roles. *Building Research & Information,* 33**,** 455-467.

CONSENSUSDOCS 2011. ConsensusDocs™ 301 Building Information Modelling (BIM) addendum.

COSTA, D. B., FORMOSO, C. T., KAGIOGLOU, M., ALARCÓN, L. F. & CALDAS, C. H. 2006. Benchmarking Initiatives in the Construction Industry: Lessons Learned and Improvement Opportunities. *Journal of Management in Engineering,* 22.

CROSTHWAITE, D. 2012. World Construction 2012. An AECOM Company.

CUI, X. 2012. *The Study of BIM Application Maturity Model.* MsC, Harbin Institute of Technology.

DANTZIG, G. B. 1963. Linear programming and extensions. NJ: Princeton University Press: Princeton University.

DAVENPORT, T. H. 1992. *Process Innovation: reengineering work through information technology,* Harvard Business Review Press.

DE MORAES, L., GARCIA, R., ENSSLIN, L., DA CONCEIÇÃO, M. J. & DE CARVALHO, S. M. 2010. The multicriteria analysis for construction of benchmarkers to support the Clinical Engineering in the Healthcare Technology Management. *European Journal of Operational Research,* 200**,** 607-615.

DEFINITION OF POLICY. 2007. *Definition of Policy, Clemson University - Office of Research Compliance* [Online]. Clemson: Clemson University. Available: http://www.clemson.edu/research/orcSite/orcIRB_DefsP.htm. [Accessed 14th April 2007].

DEHE, B. & BAMFORD, D. 2015. Development, test and comparison of two Multiple Criteria Decision Analysis (MCDA) models: A case of healthcare infrastructure location. *Expert Systems with Applications*, 42, 6717-6727.

DHURDGP 2014. BIM standard for AEC industry in Guangdong (in Chinese). Guangdong: Department of Housing and Urban-Rural Development of Guangdong Province

DIKMEN, I., BIRGONUL, M. T. & KIZILTAS, S. 2005. Prediction of Organizational Effectiveness in Construction Companies. *Journal of Construction Engineering and Management*, 131.

DING, L., ZHOU, Y. & AKINCI, B. 2014. Building Information Modeling (BIM) application framework: The process of expanding from 3D to computable nD. *Automation in Construction*.

DIXTIONARIES, O. *'Technology'* - *Compact Oxford English Dictionary* [Online]. Available: http://www.oxforddictionaries.com/definition/english/technology?q=Technology [Accessed 24th June 2015].

DOSSICK, C. S., P.E., M. A. & NEFF, G. 2010. Organizational divisions in BIM-enabled commercial construction. *JOURNAL OF CONSTRUCTION ENGINEERING AND MANAGEMENT*, 136.

DU, J., LIU, R. & ISSA, R. R. A. 2014. BIM Cloud Score: Benchmarking BIM Performance. *Journal of Construction Engineering and Management*, 140, 04014054.

DUNCAN, A. & ALDWINCKLE, G. 2015. *RE: Arup BIM Maturity Measure*

EADIE, R., BROWNE, M., ODEYINKA, H., MCKEOWN, C. & MCNIFF, S. 2013. BIM implementation throughout the UK construction project lifecycle: An analysis. *Automation in Construction*, 36, 145-151.

EASTMAN, C., TEICHOLZ, P., SACKS, R. & LISTON, K. 2011. *BIM Handbook: A Guide to Building Information Modeling for Owners, Managers, Designers, Engineers, and Contractors*, Canada, John Wiley and Sons.

EASTMAN, C. M. 1974. *An outline of the Building Description System. Research report - Institute of Physical Planning, Carnegie-Mellon University*, Institute of Physical Planning, Carnegie-Mellon University.

FISCHER, M. 2006. Formalizing Construction Knowledge for Concurrent Performance-Based Design. *EG-ICE 2006*.

FORSYTHE, P. 2014. The Case for BIM Uptake among Small Construction Contracting Businesses. *The 31st International Symposium on Automation and Robotics in Construction and Mining (ISARC 2014)*.

GAO, J. 2011. *A Characterisation Framework to Document and Compare BIM Implementations on Construction Projects.* Ph.D, Standford University.

GAO, X., ZHANG, H. & YANG, P. 2015. Study on Collaborative Design Based BIM (in chinese). *China Exploration & Design Association*, 1.

GARVIN, D. A. 1993. Building a Learning Organization. *Harvard Business Review,* 71, 78-91.

GIEL, B. 2014. MINIMUM BIM, 2nd Edition proposed revision - NBIMS v3. The Whiting-Turner Contracting Company Tamera McCuen University of Oklahoma.

GIEL, B., ISSA, R. R. A. & LIU, R. Perceptions of organizational BIM maturity variables within the US AECO industry Proceedings of the CIB W78 2012: 29th International Conference, 2012 Beirut, Lebanon.

GIEL, B. K. & ISSA, R. R. A. 2013. Return on Investment Analysis of Using Building Information Modeling in Construction. *J. Comput. Civ. Eng.,* 27, 511–521.

GOSMPG 2015. BIM standard for AEC industry in Shanghai (in Chinese). The General office of Shanghai Municipal People's Government,.

GU, N. & LONDON, K. 2010. Understanding and facilitating BIM adoption in the AEC industry. *Automation in Construction,* 19, 988-999.

GUPTA, A. 2013. *Developing a BIM-based methodology to support renewable energy assessment of buildings.* DOCTOR OF PHILOSOPHY (Ph.D), Cardiff University.

HARTMANN, T., VAN MEERVELD, H., VOSSEBELD, N. & ADRIAANSE, A. 2012. Aligning building information model tools and construction management methods. *Automation in Construction,* 22, 605-613.

HASSON, F. & KEENEY, S. 2011. Enhancing rigour in the Delphi technique research. *Technological Forecasting and Social Change,* 78, 1695-1704.

HIG 2015. The Management Maturity of AEC industry organization of China 2014 (in chinese). Headchina International Group.

HIJAZI, W., ALKASS, D. S., P.ENG. & ZAYED, D. T. 2009. Constructability Assessment Using BIM/4D CAD Simulation Model. *AACE International.*

HOWARD, R. & BJÖRK, B.-C. 2008. Building information modelling – Experts' views on standardisation and industry deployment. *Advanced Engineering Informatics,* 22, 271-280.

HSU, Y.-G., TZENG, G.-H. & SHYU, J. Z. 2003. Fuzzy multiple criteria selection of government-sponsored frontier technology R&D projects. *R&D Management,* 33.

IBRAHIM, N. H. 2013. Reviewing the evidence: use of digital collaboration technologies in major building and infrastructure projects. *Journal of Information Technology in Construction,* 18, 40-63.

INDIANA UNIVERSITY 2009b. IU BIM Proficiency Matrix.

INFOCOMM BIM TASKFORCE 2011. Building Information Modeling.

ISO STANDARD 2010. ISO 29481 - 1:2010(E): Building information modelling - Information delivery manual. Part 1: Methodology and format.

JEONG, J. S., GARCÍA-MORUNO, L. & HERNÁNDEZ-BLANCO, J. 2013. A site planning approach for rural buildings into a landscape using a spatial multi-criteria decision analysis methodology. *Land Use Policy,* 32, 108-118.

JUNG, Y. & GIBSON, E. 1999. Planning for Computer Integrated Construction. *Journal of Computing in Civil Engineering*, 13.

JUNG, Y. & JOO, M. 2011. Building information modelling (BIM) framework for practical implementation. *Automation in Construction*, 20, 126-133.

KAM, C., SENARATNA, D., MCKINNEY, B. & XIAO, Y. 2013a. The VDC Scorecard: Formulation and Validation. *CIFE Working Paper #WP135*. Center for Integrated Facility Engineering Stanford University.

KAM, C., SENARATNA, D., XIAO, Y. & MCKINNEY, B. 2013b. The VDC Scorecard: Evaluation of AEC Projects and Industry Trends. *CIFE Working Paper #WP136*. Center for Integrated Facility Engineering Stanford University.

KHOSROWSHAHI, F. & ARAYICI, Y. 2012. Roadmap for implementation of BIM in the UK construction industry. *Engineering, Construction and Architectural Management*, 19, 610-635.

KREIDER, R. G. 2011. Organisaitonal BIM Assessment. Penn State Computer Integrated Construction.

KWAK, Y. H. & IBBS, C. W. 2002. Project Management Process Maturity Model. *JOURNAL OF MANAGEMENT IN ENGINEERING*.

LEE, G. 2007. BIM collaboration methods for improving the efficiency of BIM. *BIM Workshop, Construction Association of Korea*. Seoul, Korea.

LEE, S.-I., BAE, J.-S. & CHO, Y. S. 2013. Efficiency analysis of Set-based Design with structural building information modeling (S-BIM) on high-rise building structures. *Automation in Construction*, 23, 20-32.

LEE, S., YU, J. & JEONG, D. 2015. BIM Acceptance Model in Construction Organizations. *Journal of Management in Engineering*, 31, 04014048.

LEE, W. L. & BURNETT, J. 2008. Benchmarking energy use assessment of HK-BEAM, BREEAM and LEED. *Building and Environment*, 43, 1882-1891.

LING, F. Y. Y., LOW, S. P., WANG, S. Q. & LIM, H. H. 2009. Key project management practices affecting Singaporean firms' project performance in China. *International Journal of Project Management*, 27, 59-71.

LINSTONE, H. A. & TUROFF, M. 1975. *The Delphi Method Techniques and Applications*, London, Addison-Wesley.

LONDON, K., SINGH, V., TAYLOR, C., GU, N. & BRANKOVIC, L. 2009. *Towards the development of a Project Decision Support Framework for adoption BIM*, IGI Publishing.

LOVE, P. E. D., EDWARDS, D. J., HAN, S. & GOH, Y. M. 2011. Design error reduction: toward the effective utilization of building information modeling. *Research in Engineering Design*, 22, 173-187.

LOVE, P. E. D., SIMPSON, I., HILL, A. & STANDING, C. 2013. From justification to evaluation: Building information modeling for asset owners. *Automation in Construction,* 35, 208-216.

LU, W. & OLOFSSON, T. 2014. Building information modeling and discrete event simulation: Towards an integrated framework. *Automation in Construction,* 44, 73-83.

LUU, V. T., KIM, S.-Y. & HUYNH, T.-A. 2008. Improving project management performance of large contractors using benchmarking approach. *International Journal of Project Management,* 26, 758-769.

MAHALINGAM, A., KASHYAP, R. & MAHAJAN, C. 2010. An evaluation of the applicability of 4D CAD on construction projects. *Automation in Construction,* 19, 148-159.

MARDANI, A., JUSOH, A. & ZAVADSKAS, E. K. 2015. Fuzzy multiple criteria decision-making techniques and applications – Two decades review from 1994 to 2014. *Expert Systems with Applications,* 42, 4126-4148.

MCCUEN, T. L., SUERMANN, P. C. & KROGULECKI, M. J. 2012. Evaluating Award-Winning BIM Projects Using the National Building Information Model Standard Capability Maturity Model. *JOURNAL OF MANAGEMENT IN ENGINEERING,* 28, 224-230.

MCGRAW-HILL CONSTRUCTION 2008. building information modelling: transforming design and construction to achieve greater industry productivity.

MCGRAW-HILL CONSTRUCTION 2009a. The business value of BIM: getting to the bottom line.

MCGRAW HILL CONSTRUCTION 2010. The Business Value of BIM In Europe: Getting Building Information Modelling to the Bottom Line in the United Kingdom, France and Germany

MCGRW_HILL CONSTRUCTION 2012. The Business Value in North America Multi-Year Trend Analysis and User Ratings (2007-2012).

MCIWLG 2003. Construction Information of The Next Fifteen Years' Plan (in Chinese). (in Chinese): Ministry of Construction Information Work Leading Group.

MEESAPAWONG, P. 2013. *Managing innovation in public research and development organizations using a combined Delphi and Analytic Hierarchy Process approach.* Cardiff University.

MEESAPAWONG, P., REZGUI, Y. & LI, H. 2014. Planning innovation orientation in public research and development organizations: Using a combined Delphi and Analytic Hierarchy Process approach. *Techno Fore & Social Change.*

MIETTINEN, R. & PAAVOLA, S. 2014. Beyond the BIM utopia: Approaches to the development and implementation of building information modeling. *Automation in Construction,* 43, 84-91.

MILLER, L. & LUO, H. 2015. BIM technology lead the future - interview buildingSMART director Lee Miller & Autodesk (China) technical director Haitao Luo (in chinese). *In:* ZHANG, J. & ZHAO, L. (eds.). China: Construction & design for project.

MOHURD 2013. China BIM Executive Instruction for AEC industry (in Chinese). China: Ministry of Housing and Urban-Rural Development of the People's Republic of China.

MOHURD 2014. The strategy of AEC industry development and reform. Ministry of Housing and Urban-Rural Development of the People's Republic of China.

MOHURD 2015. China BIM Executive Instruction for AEC industry - final (in Chinese). China: Ministry of Housing and Urban-Rural Development of the People's Republic of China.

MOHURD & AQSIQ Unified standard for building information model application (Draft) (in Chinese). China: Ministry of Housing and Urban-Rural Development of the People's Republic of China (MOHURD), General Administration of Quality Supervision, Inspection and Quarantine of the People's Republic of China (AQSIQ),.

MOM, M., TSAI, M.-H. & HSIEH, S.-H. 2014b. Developing critical success factors for the assessment of BIM technology adoption: Part II. Analysis and results. *Journal of the Chinese Institute of Engineers,* 37**,** 859-868.

MUTHU, S., WHITMAN, L. & CHERAGHI, S. H. Business process reengineering: a consolidated methodology Proceedings of The 4th Annual International Conference on Industrial Engineering Theory, Applications and Practice, 1999 November 17-20, 1999, San Antonio, Texas, USA.

NAJAFABADI, R. A. 2013. *An Evaluation Framework for the Integrated Design Process of Sustainable High-Performance Buildings.* Doctor of Philosophy, University of Washington.

NAPIER, B., CONNOLLY, K. J. & JERNIGAN, F. 2009. Building Information Modeling-A report on the current state of BIM technologies and recommendations for implementation. The State of Wisconsin Department of Administration Division of State Facilities.

NBS 2013a. BIM for the terrified a guide for manufacturers. Construction Products Association and NBS.

NBS 2014. National BIM Report 2014.

NEAVE, H. R. 1987. Deming's 14 Points for Management: Framework for Success. *Journal of the Royal Statistical Society,* 36.

NEDERVEEN, G. A. V. 1992. Modelling multiple views on buildings. *Automation in Construction,* 1**,** 2115-224.

NIBS 2007b. National Building Information Modeling Standard. *Version 1 - Part 1: Overview, Principles, and Methodologies.* National Institute of Building Sciences.

NIBS 2012. National BIM Standard - United States Version 2. National Institute of Building Scienses.

NIBS 2015. National BIM Standard - United States Version 3. National Institute of Building Scienses.

NYCDB 2013. Building Information Modeling Site Safety Submission Guidelines and Standards (BIM manual). New York City Department of Building.

OATES, B. J. 2006. *Researching Information Systems and Computing,* London, SAGE Publications Ltd.

ONO, R. & WEDEMEYER, D. J. 1994. Assessing the validity of the Delphi technique. *Futures,* 26, 289-304.

ONS 2010. Key Performance Indicators and Benchmarking - Construction Industry Key Performance Indicators (KPIs). Office for National Statistics.

PENN STATE UNIVERSITY 2010. BIM Project Execution Planning Guide. The Computer Integrated Construction Research Group.

PICKFORD, L. 2013. *BIM at Work: Part 1 RICS Head Office: Challenges* [Online]. isurv. Available: http://www.isurvinfo.com/papers/Incorporating-BIM-Into-Your-Business_November-2013_2.pdf?tertiary_id=32&tertiary_url=building-information-modelling&secondary_url=construction [Accessed 23rd Jan 2014].

POPOV, V., JUOCEVICIUS, V., MIGILINSKAS, D., USTINOVICHIUS, L. & MIKALAUSKAS, S. 2010. The use of a virtual building design and construction model for developing an effective project concept in 5D environment. *Automation in Construction,* 19, 357-367.

PORWAL, A. & HEWAGE, K. N. 2013. Building Information Modeling (BIM) partnering framework for public construction projects. *Automation in Construction,* 31, 204-214.

PSCIC 2013. BIM Planning Guide for Facility Owners Penn State University: Penn State Computer Integrated Construction.

REZGUI, Y., BEACH, T. & RANA, O. 2013. A governance approach for BIM management across lifecycle and supply chains using mixed-modes of information delivery. *Journal of Civil Engineering and Management,* 19, 239-258.

REZGUI, Y. & MEDJDOUB, B. 2007. *A service infrastructure to support ubiquitous engineering practices,* US, Springer.

RUSSELL, J. S. & SKIBNIEWSKI, M. J. 1987. Decision Criteria in Contractor Prequalification. *Journal of Management in Engineering,* 4.

SAATY, R. W. 1987. The analytic hierarchy process—what it is and how it is used. *Mathematical Modelling,* 9.

SEBASTIAN, R. 2011. Changing roles of the clients, architects and contractors through BIM. *Engineering, Construction and Architectural Management,* 18, 176-187.

SEBASTIAN, R. & VAN BERLO, L. 2010. Tool for Benchmarking BIM Performance of Design, Engineering and Construction Firms in The Netherlands. *Architectural Engineering and Design Management*, 6, 254-263.

SECG 2013. First Steps to BIM Competence: A Guide for Specialist Contractors. Specialist Engineering Contractors' Group.

SENATE PROPERTIES 2007. Senate Properties: BIM Requirement 2007. Finland.

SHAFIQ, M. T., MATTHEWS, J. & LOCKLEY, S. R. 2013. A study of BIM collaboration requirements and available features in existing model collaboration systems *ITCon*, 18.

SHAPIRA, A. & LYACHIN, B. 2009. Identification and Analysis of Factors Affecting Safety on Construction Sites with Tower Cranes. *Journal of Construction Engineering and Management*, 135.

SHAPIRA, A. & SIMCHA, M. 2009. AHP-Based Weighting of Factors Affecting Safety on Construction Sites with Tower Cranes. *JOURNAL OF CONSTRUCTION ENGINEERING AND MANAGEMENT*.

SHUO, W. & JIANCHENG, L. 2014. Discussion on Some Problems in the Application of the BIM in Construction Projects (in chinese). *Digital Technologies in Architecture*.

SINGH, V., GU, N. & WANG, X. 2011. A theoretical framework of a BIM-based multi-disciplinary collaboration platform. *Automation in Construction*, 20, 134-144.

SMITH, P. 2014. BIM Implementation -global strategies. *Procedia Engineering*, 85, 482-492.

SOARES, R. 2013. Reengineering Management of Construction Projects. *International Journal of Business and Social Science*, 4.

SONMEZ, M., HOLT, G. D., YANG, J. B. & GRAHAM, G. 2002. Applying Evidential Reasoning to Prequalifying Construction Contractors. *JOURNAL OF MANAGEMENT IN ENGINEERING*.

STATE OF OHIO 2010. STATE OF OHIO Building Information Modelling PROTOCOL: State Architects Offices, Geenral Services Division.

SUBRAMANIAN, N. & RAMANATHAN, R. 2012. A review of applications of Analytic Hierarchy Process in operations management. *International Journal of Production Economics*, 138, 215-241.

SUCCAR, B. 2009a. Building information modelling framework: A research and delivery foundation for industry stakeholders. *Automation in Construction*, 18, 357-375.

SUCCAR, B. 2009b. Building Information Modelling Maturity Matrix. *In:* UNDERWOOD, J. & ISIKDAG, U. (eds.) *Handbook of Research on Building Information Modeling and Construction Informatics: Concepts and Technologies.*

SUCCAR, B. 2010. The five components of BIM performance measurement. *Part of Proceedings: W096 — Special Track 18th CIB World Building Congress (combined with W104)*. Salford, UK.

SUCCAR, B. & KASSEM, M. 2015. Macro-BIM adoption: Conceptual structures. *Automation in Construction*, 57, 64-79.

SUCCAR, B., SHER, W. & WILLIAMS, A. 2012. Measuring BIM performance: Five metrics. *Architectural Engineering and Design Management*, 8, 120-142.

SUCCAR, B., SHER, W. & WILLIAMS, A. 2013. An integrated approach to BIM competency assessment, acquisition and application. *Automation in Construction*, 35, 174-189.

SUCHMAN, L. 2007. *Human–Machine Reconfigurations*, Cambridge Cambridge University Press.

SUERMANN, P. C., ISSA, R. R. A. & MCCUEN, T. L. 2008. Validation of the U.S. National Building Information Modeling Standard Interactive Capability Maturity Model. *12th International Conference on Computing In Civil and Building Engineering*. Beijing, China.

TAGCA 2006. The Contractors' Guide to BIM The Associated General Contractors of America.

TAYLOR, J. E., A.M.ASCE & BERNSTEIN, P. G. 2009. Paradigm Trajectories of Building Information Modeling Practice in Project Networks. *Journal of management in engineering* 25.

TAYLOR, J. E. & LEVITT, R. 2007. Innovation alignment and project network dynamics: An integrative model for change. *Project Management Journal*, 38, 22-35.

TEICHOLZ, P. 2013. *Labor-Productivity Declines in the Construction Industry: Causes and Remedies (Another Look)* [Online]. AECbytes. Available: http://www.aecbytes.com/viewpoint/2013/issue_67.html 2 Dec 2014].

THE CLIMATE GROUP. 2014. *CONSTRUCTION WASTE RECYCLING IN CHINA: THE CLIMATE GROUP RELEASES NEW REPORT* [Online]. Available: http://www.theclimategroup.org/what-we-do/news-and-blogs/construction-waste-recycling-in-china-the-climate-group-releases-new-report/ [Accessed 15th November 2014].

TOLMAN, F. P. 1999. Product modelling standards for the building and construction industry: past, present and future. *Automation in Construction*, 8, 227-235.

TRIANTAPHYLLOU, E. 1997. A Sensitivity Analysis Approach For Some Deterministic Multi-Criteria Decision Making methods. *Decision Sciences*, 28.

TSAI, M.-H., KANG, S.-C. & HSIEH, S.-H. 2010. A three-stage framework for introducing a 4D tool in large consulting firms. *Advanced Engineering Informatics*, 24, 476-489.

TSAI, M.-H., KANG, S.-C. & HSIEH, S.-H. 2014. Lessons learnt from customization of a BIM tool for a design-build company. *Journal of the Chinese Institute of Engineers*, 37, 189-199.

TSAI, M.-H., KANG, S.-C. & HSIEH, S.-H. 2014c. Lessons learnt from customization of a BIM tool for a design-build company. *Journal of the Chinese Institute of Engineers*, 37.

TSAI, M.-H., MOM, M. & HSIEH, S.-H. 2014a. Developing critical success factors for the assessment of BIM technology adoption: part I. Methodology and survey. *Journal of the Chinese Institute of Engineers*, 37, 845-858.

UPSEIS. *How Are Earthquake Magnitudes Measured?* [Online]. Available: http://www.geo.mtu.edu/UPSeis/intensity.html [Accessed 22nd Septermber 2015].

VAIDYA, O. S. & KUMAR, S. 2006. Analytic hierarchy process: An overview of applications. *European Journal of Operational Research*, 169, 1-29.

WANG, K., WANG, C. K. & HU, C. 2005. Analytic Hierarchy Process With Fuzzy Scoring in Evaluating Multidisciplinary R&D Projects in China. *IEEE TRANSACTIONS ON ENGINEERING MANAGEMENT*, 52.

WANG, X., MANCINI, M., NEPAL, M., CHONG, H.-Y., SKITMORE, M. & ISSA, R. 2015. Call for Papers. *International Journal of Project Management*, 33, 479-480.

WEGELIUS-LEHTONEN, T. 2001. Performance measurement in construction logistics. *Int. J. Production Economics*, 69, 107-116.

WICKS, A. C. & FREEMAN, R. E. 1998. Organization Studies and the New Pragmatism: Positivism, Anti-positivism, and the Search for Ethics. *Organization Science*, 9, 123-140.

WON, J., LEE, G., DOSSICK, C. & MESSNER, J. 2013. Where to focus for successful adoption of building information modeling within organization. *Journal of Construction Engineering and Management*, 11.

WONG, J. K.-W. & KUAN, K.-L. 2014. Implementing 'BEAM Plus' for BIM-based sustainability analysis. *Automation in Construction*, 44, 163-175.

WSP 2011. Truths of BIM.

XU, D. & YANG, J. 2003. Intelligent Decision Systemfor Self-Assessment. *J. Multi-Crit. Decis. Anal.*, 12, 43-60.

YANG, J. 2001. Rule and utility based evidential reasoning approach for multiattribute decision analysis under uncertainties. *European journal of operational research*, 131, 31-61.

YU, I., KIM, K., JUNG, Y. & CHIN, S. 2007. Comparable performance measurement system for construction companies. *JOURNAL OF MANAGEMENT IN ENGINEERING*, 23, 131-139.

ZAHRIZAN, Z., ALI, N. M., HARON, A. T., MARSHALL-PONTING, A., ABD, Z. & HAMID 2013. Exploring the Adoption of Building Information Modelling (BIM) in the Malaysian Construction Industry: A Qualitative approach. *International Journal of Research in Engineering and Technology*, 2.

ZHU, B. & XU, Z. 2014. Analytic hierarchy process-hesitant group decision making. *European Journal of Operational Research*, 239, 794-801.

ZHU, L., AURUM, A. K., GORTON, I. & JEFFERY, R. 2005. Tradeoff and Sensitivity Analysis in Software Architecture Evaluation Using Analytical Hierarchy Process. *Software Quality Journal*, 13.

POSTSCRIPT

The author, Dr. Keyu Chen has been in the Architecture, Engineering and Construction industry since 2005. He has a great interest in the building industry: how the building was designed, how the crane was set up, how everything was constructed and managed. Singapore as one of the most developed countries in Asia has a number of high-rise iconic buildings, thus he chose this multicultural country as the first step to his further education, as well as career. He started his journey by studying in Singapore Polytechnic (SP) in Civil and Structure Engineering. The first game changing technology he was working on for his Final Year Dissertation is Virtual Reality (VR), which is still one of the advancing technologies even today.

After the author graduated from SP, he worked in Bachy Soletanche Singapore Branch on the MRT Circle Line project. As a modeler and design coordinator, he realized that 3D visualization and project integration will play a critical role in terms of productivity. Shortly after taking on this role, he decided to continue his education in Cardiff University, one of the best Civil Engineering Programmes in the UK, to gain more knowledge in this field.

By the end of his degree of study, the concept: Building Information Modelling – BIM was raised. The author realized this as a mechanism to tackle the problems he had faced in his work at Bachy Soletanche.

Dr. Haijiang Li, took the author further in to world of the BIM through working together with him on his degree dissertation: BIM in construction and coordination. However, the author believes BIM should get rolled in a much earlier phase with a more strategic manner to enhance its effectiveness along the entire asset's lifecycle. Dr. Haijiang Li helped discover the author's passion in BIM and as a result of his successful dissertation, offered the author an opportunity in a Ph.D. position.

During the author's Ph.D., he had been involved in quite a few BIM based UK industry projects and R&D projects. He was also hired as a BIM manager to guide one of the design firms in China for BIM adoption on a real project with positive outcome. He had tested his knowledge in BIM as well as absorbed new knowledge and experience.

After the author graduated from his Ph.D., he worked for Arcadis in the UK. He worked with an excellent team on a core role: BIM strategy development. Whilst working as a BIM coordinator the author was also furthering his experience as a civil designer and a project management assistant. All these valuable working experiences have provided the author with a very good fundamental knowledge in the Architecture, Engineering and Construction industry.

In order to identify how BIM can be promoted from the government's perspective, the author had decided to move back to Singapore through the Building Construction Authority (BCA) of Singapore had made a lot of effort to push BIM in Singapore's industry. As a civil servant in Building and Construction Authority (BCA) in Singapore, the author's main roles are: Programme Manager and teacher for Construction and Coordination for a Specialist Diploma in BIM – a government recognized BIM certificate for Adult Learning. In addition, he is in charge of BIM and other IT programme leaders for the Full Time Diploma students and provides lecture and laboratory applications for these courses. At the moment, the author is working on the BIM Guide for Asset Information Delivery.

The author believes BIM has a great potential to improve the whole AEC industry's productivity, with the right push. He believes that this strategy is needed from a regulatory body's perspective to control an individual's BIM capability: to align with the objectives and demands of the industry, so as to influence the overall performance related to BIM. The author's second book which addresses these matters and additional concepts is under preparation and will be published soon.

The author is also dedicated to charity work. He is currently working towards establishing his own charity organization. Part of the profit made from the book sale will be donated to this organization for charity purposes. Alternatively, if one wishes to make a donation, these will also be kindly accepted. The ultimate objective is to provide an equal environment to those disabled members of the community who are struggling to get jobs due to their disabilities.

For more information, please email the author: Chen.ky@outlook.com

This book is under the supervision of Dr. Haijiang Li

Dr. HaiJiang LI
Reader in Engineering Informatics
Director of Cardiff BIM and VR lab
FBCS – Fellow of British Computer Society
Secretary of European Group of intelligent computing
Technical board member of BuildingSMART UK
Visiting Professor for Dalian University of Technology, China
External BIM Expert for China State Construction LTD
External BIM Expert for China Communication Construction LTD
External BIM Expert for Lubansoft BIM Institution

For more than 20 years, my research and industry practice have been interfacing between computing (HPC / Grid / Cloud computing, Big Data, Artificial Intelligence - ontology, sensors, automatic control, multi-agent system, BIM, immersive virtual reality etc.) and different engineering domains / sectors via design, simulation, optimization, visualization and management, e.g. civil and hydraulic engineering – water and environment issues; structural engineering and asset management for building and infrastructure – road, bridge, harbour and dam – structural design and analysis, onsite construction and project management; sustainable engineering - energy saving in building and district level, resilience and disaster management for built environment.

As a Principle / Co-Investigator, the total value of my research projects is in excess of £30M (about £5M as PI); More than 120 publications, about half are peer reviewed journal / IEEE papers; Head of Cardiff BIM research group and its associated BIM & VR (Virtual Reality) laboratory, with a team of 23 PhDs (11 graduated) and 10 research associates / fellows; Chairing / attending major international conferences (more than 40 cases of invited speeches, including 12 keynotes), editorial board member for 4 international journals; reviewer for EPSRC, NSERC (Canada) and for 41 journals and conferences, editing (as guest editor) special issues.

www.ingramcontent.com/pod-product-compliance
Lightning Source LLC
Chambersburg PA
CBHW040314220526
45473CB00009B/2425